Rodents are the predominant experimental animals found in life-sciences research laboratories. The body temperature of a rodent is markedly affected by surgical, chemical, or environmental manipulation. Because temperature regulation is controlled essentially by a "holistic" regulatory system, meaning that its responses affect the activities of all other physiological and behavioral processes, it is clear that researchers working with rodents must be familiar with thermoregulatory physiology.

With the help of extensive data tables and figures, this book explains the key facets of rodent thermal physiology, including neurological control, metabolism, thermoregulatory effectors, core and brain temperatures, circadian rhythm, developmental patterns and aging, temperature acclimation, and gender and intraspecies variations. There is a novel chapter on the effects of trauma, toxic chemicals, and other factors. Mouse, gerbil, hamster, rat, and guinea pig are the rodents discussed. The book should therefore find use in government, academic, or industrial laboratories whose researchers are working with rodents.

Temperature regulation in laboratory rodents

Temperature regulation in laboratory rodents

CHRISTOPHER J. GORDON

Research Triangle Park, North Carolina

CAMBRIDGE
UNIVERSITY PRESS

Published by the Press Syndicate of the University of Cambridge
The Pitt Building, Trumpington Street, Cambridge CB2 1RP
40 West 20th Street, New York, NY 10011-4211, USA
10 Stamford Road, Oakleigh, Victoria 3166, Australia

First published 1993

Printed in the United States of America

Library of Congress Cataloging-in-Publication Data
Gordon, Christopher J.
Temperature regulation in laboratory rodents /
Christopher J. Gordon.
p.   cm.
Includes bibliographical references.
ISBN 0-521-41426-1 (hardback)
1. Body temperature – Regulation.   2. Physiology, Comparative.
I.   Title.
[DNLM: 1. Animals, Laboratory.   2. Body Temperature Regulation –
physiology.   3. Physiology, Comparative.   4. Rodentia – physiology.
QY 60.R6 G662c]
QP135.G57   1993
619'.93 – dc20
DNLM/DLC   92-49550
for Library of Congress   CIP

A catalog record for this book is available from the British Library.

ISBN 0-521-41426-1 hardback

This book was written by Christopher J. Gordon in his private capacity.
No official support or endorsement by the Environmental Protection
Agency or any other agency of the federal government is intended or
should be inferred.

To my loving wife, Susie,
and children, Kevin and Karen

Contents

Preface

Why do we need a book devoted to the thermoregulatory characteristics of laboratory rodents? Such a book will be of obvious benefit to those involved in the study of thermal biology, but this book is also written to meet the needs of more readers than the relative handful of thermal biologists. One or more species of laboratory rodents are used predominantly by researchers in a variety of fields in the life sciences, including neural science, endocrinology, immunology, nutrition, and many others. The biological endpoints commonly measured in these fields would seem not to be related to thermoregulation. Yet manipulation of any one of these systems with surgical, pharmacological, and/or environmental procedures often leads to changes in the rodent's thermoregulation. Because temperature regulation is controlled essentially by a "holistic" regulatory system, meaning that its responses affect the activities of all other physiological and behavioral processes, it is clear that researchers working with rodents must be cognizant of thermoregulatory physiology.

Since completing my graduate work, I have found a need for a comprehensive source on the thermal physiology of laboratory rodents. Most other books on temperature regulation have focused on specific aspects of thermoregulation, such as fever, pharmacological control, exercise, and nonshivering thermogenesis. Other books have concentrated on specific mammalian species, particularly humans and the domesticated species that are of importance to agriculture. However, there are few sources that have examined the responses of a specific group of mammals such as the rodents. In 1971 the eminent Dr. J. S. Hart prepared a thorough monograph on the thermoregulatory characteristics of both wild and domesticated rodents (Hart, 1971). Although that is an excellent source for most thermal biologists, I have often thought that it does not necessarily address the needs of researchers whose primary interests are in fields other than thermoregulation. To address these needs, in 1990 I wrote a review on the thermal biology of the laboratory rat

(Gordon, 1990a). The popularity of that review, as judged from the number of reprint requests, was considerable. Because of the strong response to that review and encouragement from the editors of Cambridge University Press, I was convinced that a thorough comparative analysis of thermoregulatory responses in the commonly used laboratory rodents would be most useful.

The style of this book differs somewhat from the norm. I believe that a comparative book calls for a thorough presentation of the pertinent data from the literature. Because of the variability in some thermoregulatory parameters in a given species, I found it necessary to tabulate data from a number of laboratories to facilitate comparison. The tables allow the reader to quickly assess and compare the thermoregulatory characteristics of a particular species. Moreover, it is hoped that voids in the data presented in these tables will indicate where future research should be directed. In preparing the illustrations, I sought, whenever possible, to show the responses of two or more rodent species. The level of detail in this book was one of my greatest concerns. With over 700 references, it should be obvious that I have tried to make the coverage as extensive as possible without exceeding the publisher's space limitation. However, in the course of doing the research for this book it is likely that I overlooked some pertinent papers. To the authors of those papers I apologize, and I hope that they will inform me of my oversights.

I have attempted to write this text at a level that should meet the needs of researchers with a minimum background of upper-level undergraduate courses in physiology and/or related fields. The book covers the physiological and behavioral responses of the principal laboratory rodents: mouse, gerbil, hamster, rat, and guinea pig. Although recent molecular studies have suggested that the guinea pig is not a true rodent, its thermoregulatory responses are considered here along with those of other rodent species. What I, and many other researchers, consider to be the most pertinent facets of thermoregulatory physiology are covered: neurological control, metabolism, homeostasis of body and brain temperatures, stress, motor effectors, growth and development, temperature acclimation, and intraspecies and gender differences. The final chapter deals with a novel aspect not commonly presented in other texts, but nonetheless crucial to today's researcher: the effects of adverse perturbations, including hypoxia, chemical toxicity, and trauma, on thermoregulation.

Acknowledgments

This book would not have been possible if not for the invaluable training and guidance I received from my postgraduate advisors: Dr. J. Homer Ferguson, formerly of the University of Idaho, and Dr. James E. Heath of the University of Illinois, Urbana-Champaign. They have been inspirational in stimulating my interest in environmental and comparative physiology. I am appreciative of the efforts of the following colleagues who graciously provided critiques of chapters in the book: Drs. W. C. Duncan, A. H. Rezvani, D. B. Miller, W. P. Watkinson, R. Refinetti, S. C. Wood, and E. Berman. Although they are too numerous to name, I would like to thank all the thermal physiologists who kindly furnished me with up-to-date papers from their laboratories for the preparation of the book. I am also grateful to Drs. K. Zylan and R. Smith of Cambridge University Press for providing invaluable editorial assistance. Finally, I thank all the members of my family for their abiding support in this endeavor.

List of abbreviations

ATP, adenosine triphosphate
AVA, arterial-venous anastomosis
AVP, arginine vasopressin
BAT, brown adipose tissue
BBB, blood-brain barrier
BMR, basal metabolic rate
C', whole-body thermal conductance
CIVD, cold-induced vasodilation
CNS, central nervous system
CT_{max}, critical thermal maximum
CT_{min}, critical thermal minimum
CTR, circadian temperature rhythm
DIT, diet-induced thermogenesis
ECG, electrocardiogram
EDT, elevated defended temperature
EEG, electroencephalogram
EHL, evaporative heat loss
EMG, electromyogram
EP, endogenous pyrogen
EWL, evaporative water loss
F344, Fischer 344
FFAs, free fatty acids
FSH, follicle-stimulating hormone
GDP, guanosine diphosphate
IL-1, interleukin 1
IR, infrared
LCT, lower critical ambient temperature
LD_{50}, lethal dose for 50% mortality
LE, Long-Evans

LH, luteotropic hormone
LOMR, least observed metabolic rate
LPS, lipopolysaccharide
M (MR), metabolic rate
MAP, mean arterial pressure
MMR, maximum metabolic rate
MOMR, minimum observed metabolic rate
NE, norepinephrine (noradrenaline)
NPRQ, nonprotein respiratory quotient
NST, nonshivering thermogenesis
OVLT, organovasculosum lamina terminalis
PGE_n, prostaglandin E_1 or E_2
PMR, peak metabolic rate
POAH, preoptic area/anterior hypothalamus
PS, paradoxical sleep
PVMT, peripheral vasomotor tone
RF, radio frequency
RMR, resting metabolic rate
RQ, respiratory quotient
SA, surface area
SCN, suprachiasmatic nucleus
SD, Sprague-Dawley
SHR, spontaneously hypertensive rat
SMR, standard metabolic rate
SWS, slow-wave sleep

Selected T_a, selected ambient temperature

T_a, ambient temperature

TBTS, total-body thermosensitivity

T_b, body temperature (core temperature)

T_c, core temperature

T_h, hypothalamic temperature

T_{set}, set-point temperature

T_3, triiodothyronine

T_4, thyroxine

T5′ D, thyroxine 5′-deiodinase

TNZ, thermoneutral zone

UCP, uncoupling protein

UCT, upper critical ambient temperature

WAT, white adipose tissue

1

Introduction to temperature regulation

Temperature regulation (or thermoregulation) can be defined as the control of the temperature(s) of a body under finite environmental conditions. Regulation is achieved by controlling heat gain and heat loss between the body and the environment through the utilization of autonomic and behavioral mechanisms. Birds and mammals have evolved a battery of behavioral and autonomic motor outputs (i.e., effectors) to regulate their core body temperatures within narrow limits when subjected to a wide range of ambient temperatures (Prosser and Heath, 1991). In some cases, reptiles, fish, and amphibians are able to regulate their body temperatures by means of behavioral responses. Invertebrates are *temperature conformers,* meaning that their body temperatures usually are about the same as that of their surrounding environment. Even the most primitive organisms display thermotropism (i.e., the tendency to turn toward or away from a heat source), and many temperature conformers have distinct behavioral thermoregulatory responses (Whittow, 1970; Prosser, 1973). It should be remembered that ambient temperature is probably the most critical environmental factor in limiting an organism's choices among possible habitats. Thus the development of temperature regulation undoubtedly has played a major role in evolution.

1.1. A brief historical perspective

The existence of thermoregulatory systems probably was one of the earliest discoveries of an involuntary homeostatic process, as reviewed by Lomax (1979), Folk (1974), and Hensel (1981). That is, before recorded history humans surely must have recognized that many diseases were associated with a feeling of warmth on the skin, whereas death brought on a loss of heat. The ancient Greek philosophers conceived a relationship between body heat and the vitality of living organisms. They believed that the left ventricle was the source of one's innate heat, and respiration was a vital process needed to cool the body.

Basic empirical observations mixed with religious dogma were used to explain thermoregulatory phenomena in humans until late in the seventeenth century. It was not until the late eighteenth century that modern theories of thermoregulation began to take shape. Lavoisier made some of the first measurements of heat loss in a rat, using a crude, albeit accurate, ice-bath calorimeter. In the early 1800s, measurements of body temperature and cold resistance in various species of mammals and birds were undertaken.

The development of the clinical thermometer by Allbutt in 1867 spurred research on fever and other thermoregulatory processes. Indeed, recognition of the existence of fever was perhaps the most critical factor in galvanizing research into the mechanisms of temperature regulation. During the latter 1800s it was realized by Libermeister and others that fever was not the cause of disease, but rather was a regulated elevation in core temperature and a symptom of disease.

By the early 1900s, as a result of the pioneering work of Rubner and others using direct calorimeters, a substantial data base on body temperatures and metabolism among numerous species had been developed. Detailed monographs on the temperature regulation of the laboratory rat had been published by the 1930s (e.g., Benedict and MacLeod, 1929). One of the most exciting areas of investigation into temperature regulation during the first half of the twentieth century was the search for the central nervous system (CNS) loci involved in the control of body temperature. Researchers such as Bazett, Ranson, and Magoun in the 1930s were instrumental in developing the concept of the anterior/posterior hypothalamus as the key regulatory site for control of body temperature. Spurred by military interests, research in the 1940s and 1950s focused on understanding the limits of thermoregulation in humans and other species when exposed to severely warm and cold environments. Also during that time, considerable progress was made in measuring thermal homeostatic processes in reptiles and other lower vertebrate species (e.g., Cowles and Bogert, 1944).

The work by Irving and Scholander was particularly noteworthy in describing the relationship between a species' thermoneutral profile and its adaptability to cold environments (e.g., Scholander et al., 1950). Soon after that, a multitude of studies on the thermoregulatory characteristics of wild and domesticated rodent species were carried out by J. S. Hart, F. Depocas, and many others (e.g., Hart, 1971). In the 1960s, the utilization of stereotaxically implanted thermodes to heat and cool the brain stem in awake animals (e.g., Hammel, 1968; Heath, Williams, and Mills, 1971), the first recordings of hypothalamic temperature-sensitive neurons by Nakayama (Boulant and Dean, 1986), and the discovery that body temperature could be manipulated by CNS injections of neurotransmitters (Feldberg and Myers, 1964) elevated thermoregulatory research in experimental animals and humans to a new level of understanding. Another key advancement in the field was the development of a concise glossary of thermal physiology, first prepared by Bligh and Johnson in 1973. That glossary was later expanded and refined by the International Union of Physiological Sciences (IUPS, 1987).

1.2. Current research status of thermoregulation

The past status and current status of thermoregulatory research in mammals pose somewhat of a paradox. On the one hand, as can be seen from the foregoing discussion, thermoregulatory research has played a key role in the development of our modern concepts of physiology. Yet, of the thousands of researchers in the life sciences and related fields, only a relative handful currently dedicate their primary research to thermoregulation, totaling approximately 300 worldwide (Refinetti, 1990a).

A variety of reasons can be cited for the waning interest in basic thermoregulatory research. In my opinion, a chief factor is that because the thermoregulatory system works so well, it has never been seen as a prime area in need of research funding. That is, thermal homeostasis in humans and other mammals is maintained 24 hr per day from soon after birth until death, with relatively little aberration. Congenital defects in thermoregulation in humans are extremely rare. Moreover, under most environmental conditions (i.e., barring extreme thermal stress, drug overdose, severe trauma, etc.), thermoregulatory systems rarely experience sudden failure or show "life-threatening" deficits. On the other hand, dysfunctions of other regulatory systems of key interest in the biomedical community, such as the cardiovascular, renal, hepatic, gastrointestinal, and immune systems, are the major causes of human disease and mortality. Indeed, often medications must be prescribed for much of person's lifetime to prevent malfunction or failure of the heart, kidney, liver, CNS, or other systems. But thermoregulation generally operates with few flaws until one is near the point of death, and therefore it has not been a major focus in the funding decisions in biomedical and related research in recent times.

After reading the foregoing, one might view the future of thermoregulatory research with pessimism. There are, nonetheless, several reasons that research in this field is essential:

1. That the thermoregulatory system works so well clearly makes it an excellent paradigm of biological regulation. The system achieves a fine degree of control by utilizing the functions of other organ systems for its motor outputs: the respiratory and digestive systems (e.g., salivary glands in rodents) for evaporation, the cardiovascular system for skin temperature, and skeletal muscle for shivering and behavioral thermoregulatory responses. Only brown adipose tissue appears to have evolved solely for a thermoregulatory function. Surely, if one could fully understand the workings of a multiorgan regulatory system that rarely fails, then a better understanding of other, more vulnerable autonomic systems should be forthcoming.

   Of course, the thermoregulatory system is not flawless. Like other bodily functions, thermoregulation is most susceptible to dysfunction during the early and late stages of life. The very young and the elderly are more likely to encounter thermoregulation-related maladies such as hypothermia and hyperthermia. Certain pharmaceutical therapies, toxic chemicals, stress, and other environmental and biological agents can reduce the normal ambient temperature range of thermoregulation and thereby increase its susceptibility to dysfunction.

2. Like the output from other autonomic systems, the output of the thermoregulatory system affects the functioning of all other physiological processes. That is, mammalian enzymes have evolved in a relatively stable thermal environment of 37°C and have narrow temperature ranges for optimal functioning. Thus, deviations from normal body temperatures will lead to significant changes in enzymatic activity and, hence, the functioning of a given physiological system. This becomes especially pertinent in rodents, which have more labile thermoregulatory systems than do larger species (see Chapter 9).

3. The thermoregulatory system is unique among the homeostatic processes from the standpoint that it relies on higher-level CNS processes for thermal detection and elicitation of corrective behavioral motor responses. Behavioral sensation of skin temperature is an incredibly sensitive and continuous process. It is of paramount importance in the activation of autonomic and behavioral motor outputs. Yet, in the absence of behavioral responses, such as during sleep and certain forms of anesthesia, thermoregulation is still operative. On the other hand, behavioral input–output processes are not as pivotal in other autonomic processes, such as the regulation of blood pressure and of electrolyte and water balances.

The behavioral sensitivity of thermoregulation can be exemplified by a simple examination of human energy usage. Although autonomic motor outputs can achieve a regulated core temperature over a relatively wide range of ambient temperatures, our behavioral thermoregulatory processes cause us to expend billions of kilowatt-hours (and dollars!) in heating and air conditioning to keep our environment at a comfortable temperature of 22–26°C (72–78°F).

Behavioral thermoregulation is also critical in the thermoregulation of rodents (see Chapter 4). Behavioral responses are severely compromised when restraint is used, and totally abolished with anesthesia. Although these procedures are common to many experimental designs, their use in studying thermoregulation will eliminate a significant facet of the animal's motor outputs.

4. The process of fever, which involves a unique integration of the immune and thermoregulatory systems, continues to be a critical aspect of biomedical research. There still is considerable debate over the role of an elevated body temperature during fever and whether or not it is beneficial to recovery from infection. Elucidating the mechanisms of fever will continue to be a major endeavor in thermal physiology.

5. Manipulating the output of the thermoregulatory system (e.g., raising or lowering the core temperature) has been and will continue to be a crucial aspect of various surgical and therapeutic procedures. For example, without the advantage of hypothermia to lower the body's oxygen demand, it would be impossible to perform many of today's common surgical procedures on the heart and brain. Forced elevations and reductions in body temperature have proved to be extremely beneficial in treating some maladies, including some types of cancer and trauma to the CNS. Moreover, the efficacy and toxicity of drugs and other chemical agents can be markedly altered by changes in body temperature. This point is extremely important in rodent studies, because their body temperatures can change relatively quickly when subjected to a variety of stresses and trauma (see Chapter 9).

1.3. Why study laboratory rodents?

On the basis of these five points one can infer that there are two main avenues of thermoregulatory research: (1) studies on basic thermoregulatory mechanisms, including neural physiology, comparative differences, and ecological relationships, and (2) multidisciplinary research, in which the activity of the thermoregulatory system may affect the functioning of another system that is the primary interest of the investigator. From this concern, a question arises: Which species is most appropriate for a given line of research in pure thermoregulatory research and/or a related response? The answer will depend on many factors, such as a species' thermoregulatory qualities per se, as well as its nutritional, cardiovascular, neural, immunological, and other characteristics that may be pertinent to one's specific line of research. At this point, a quotation from the eminent comparative physiologist C. Ladd Prosser is appropriate:

> Comparative physiology differs from other kinds of physiology in that the comparative approach uses the kind of organism as an experimental variable, and it emphasizes the long evolutionary history to life in diverse environments. [*Annu. Rev. Physiol.*, 48:1–6, 1986]

In other words, in the selection of a particular species of rodent or other organism for study, the researcher should be cognizant of its unique characteristics. Such knowledge can be invaluable for understanding the mechanisms of operation of a physiological system. This book strives to present these unique characteristics of the thermoregulatory systems of laboratory rodents in the hope of facilitating the comparative approach.

It is interesting to note that the amount of thermoregulatory research with rats and mice has increased considerably since around 1980 (Figure 1.1). The numbers of studies with hamsters and guinea pigs have been relatively constant since 1970, whereas work with gerbils has contributed a relatively small portion to the total rodent research effort. Research in nonrodent species, including cats, dogs, rabbits, and nonhuman primates, has shown a downward trend (Figure 1.1). There are perhaps several reasons for these patterns in animal research. The foremost factor obviously is the soaring cost of such research, which has led many to work with the less expensive laboratory rodents. Moreover, continuing pressure from various antivivisection groups has unfortunately caused many researchers to abandon their work with species such as cat, dog, rabbit, and monkey in favor of rodents.

1.4. Overview of temperature regulation in rodents

1.4.1. Terminology

The studies of temperature regulation can be broadly divided into two groups, dealing with species that are *tachymetabolic* and those that are *bradymetabolic*. Tachymetabolic species, which include birds and mammals, have relatively high basal metabolic rates compared with bradymetabolic species,

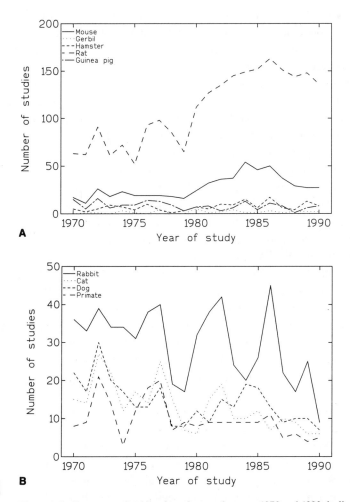

Figure 1.1. Frequency of publication of papers between 1970 and 1990 dealing with temperature regulation in key rodent (A) and nonrodent (B) species. These data were collected using a Medline literature search. Because of the restricted number of journals and possible ambiguity in the original indexing of titles, these graphs are likely to underestimate the actual numbers of studies.

including reptiles, amphibians, fish, and other species. Tachymetabolic species are also *endothermic,* meaning that their control of body temperature is dependent primarily on the generation of heat through metabolic processes. On the other hand, bradymetabolic species are *ectothermic,* meaning that their regulation is achieved behaviorally by controlling the transfer of heat between their bodies and their environment.

As with any scientific discipline, the terminology of temperature regulation can sometimes be ambiguous, depending on the species and the environmental circumstances (IUPS, 1987). For example, an endothermic species is almost always classified as a homeotherm, meaning that it can regulate its core temperature within relatively narrow limits (ca. ± 2°C). Yet many endotherms are not always homeothermic. Some birds and small mammals undergo diurnal or nocturnal torpor or annual periods of hibernation, allowing their core temperatures to drop markedly below the normal limits of homeothermy. In these instances the term *heterothermy* is used, wherein tachymetabolic species show marked daily or annual changes in core temperature.

Likewise, ectotherms are generally poikilothermic, meaning that in the absence of behavioral adjustments the animal's core temperature is closely dependent on the ambient temperature. However, many ectotherms, especially reptiles, use behavior to regulate body temperature within relatively narrow limits and thus display homeothermic characteristics. Also, some species of insects and fish, though classified as bradymetabolic, are nonetheless capable of endothermy and can actively increase their internal temperatures well above ambient levels (Prosser, 1973; Schmidt-Nielson, 1975b).

To summarize, the rodents are all tachymetabolic and rely on endothermy to regulate their core temperatures within relatively narrow limits. Among the laboratory rodents to be discussed in this book, the rat (*Rattus norvegicus*) and guinea pig (*Cavia porcellus*) are continuously homeothermic and are incapable of lowering their body temperatures under most environmental circumstances. The mouse (*Mus musculus*) is also homeothermic most of the time, but is capable of undergoing torpor during periods of food deprivation and thus can be classified as a heterotherm. The golden or Syrian hamster (*Mesocricetus auratus*) is also heterothermic at times and is capable of hibernating under proper environmental circumstances. The Mongolian gerbil (*Meriones unguiculatus*) is homeothermic most of the time but is apparently capable of torpor. It should be noted that this species has not been as well studied as other rodents.

1.4.2. Heat balance

The body's heat-balance equation is an appropriate place to begin a discussion of the fundamentals of temperature regulation in rodents and other species. It is a mathematical expression derived from the first law of thermodynamics, and it relates metabolic rate, work, and the four avenues for heat exchange to a species' bodily heat balance (IUPS, 1987):

$$S = M - W - E - C - K - R \tag{1.1}$$

where S is heat storage in the body (positive for an increase), M is metabolic rate, W is work (positive for mechanical work transferred to the environment, negative when mechanical energy is transferred from the environment to the body and is eventually converted to heat), E is evaporative heat transfer, C is convective heat transfer, K is conductive heat transfer, and R is radiative heat transfer. Each avenue of heat transfer is positive when there is a net loss of heat from the subject to the environment, and negative with heat gain. All variables in equation (1.1) are generally in standard energy units: watts (W), watts per square meter (W m^{-2}), or watts per kilogram (W kg^{-1}).

The avenues of heat exchange are factors of greater or lesser importance depending on the species and the environmental conditions. The rate of conductive heat transfer usually is quite low because so little of the animal's bare surface comes in direct contact with the substrate, but the rate of conductive heat transfer can become very high during water immersion. The rate of evaporative heat transfer is quite low under standard, room-temperature conditions (ca. 20–22°C), accounting for approximately 20% of total heat loss. Evaporative heat transfer increases markedly as ambient temperature is elevated above thermoneutrality (see Chapter 4). Thus, most of an animal's metabolic heat is dissipated by way of radiation and convection. Convective heat transfer increases in proportion to wind velocity, but air movement is generally quite minimal in most laboratory situations. Unfortunately, relatively little work has been done in laboratory rodents to measure the partitioning of heat exchange through the four major avenues, as compared with the number of studies in humans and other large species (e.g., Monteith and Mount, 1974).

Heat storage usually is expressed in terms of the rate of change in stored heat in the body and can be calculated as

$$S \text{ (W)} = \frac{c(\overline{T}_{b_1} - \overline{T}_{b_2}) \text{ body weight (g)}}{\text{time}_2 - \text{time}_1 \text{ (sec)}} \tag{1.2}$$

where c is the specific heat of the tissues ($\sim$3.47 J g^{-1} °C^{-1}), and $\overline{T}_{b1}$ and $\overline{T}_{b2}$ are the mean body temperatures at the beginning and end of the time period. Thus, under conditions where heat production is equal to the sum of all avenues of heat loss, S is equal to zero, and the animal is normothermic (synonyms: cenothermic or euthermic); see IUPS (1987). When heat production exceeds heat loss, such as during exercise or following administration of a drug that stimulates cellular metabolism, S is positive, and the animal is hyperthermic. On the other hand, when heat loss exceeds heat production, such as during acute cold exposure or following administration of a drug that induces peripheral vasodilation, S is negative, and the animal is hypothermic.

When body temperature is constant (i.e., $S = 0$) and no work is being done, an endotherm's metabolic rate (M, in watts) must therefore match its total rate of heat loss (H_t) to the environment:

$$H_t = M = R + K + C + E \qquad (1.3)$$

Incorporating specific terms for each of the avenues of heat loss, the metabolic rate is then calculated as

$$M = \epsilon\sigma A_r(T_b^4 - T_a^4) + h_k A_k(T_b - T_a) + h_c A_c(T_b - T_a) + \lambda\text{EWL} \quad (1.4)$$

where ϵ is the emissivity, or the ratio of radiant energy emitted by a body to the energy emitted by a full radiator at the same temperature (the value of ϵ is generally assumed to be 1.0 for animals, σ is the Stefan-Boltzmann constant (5.67×10^{-8} W m^{-2} °K^{-4}), A_r is the effective surface area for radiative heat exchange, h_k is the thermal-conductivity coefficient, A_k is the effective surface area for conductive heat exchange, h_c is the convective heat-exchange coefficient, A_c is the effective surface area for convective heat exchange, λ is the latent heat of vaporization (e.g., 2,411.3 J g^{-1} at 34°C), and EWL is the rate of evaporation of water. T_b is body temperature, but most appropriately would be the surface temperature of the skin or fur, and T_a is ambient or air temperature. (Note: For the radiative terms, T_a and T_b are in degrees Kelvin.)

There are several precautions to be considered when applying equation (1.4) to rodent thermal physiological studies. T_a is assumed to equal air temperature only when the temperature of the substrate (e.g., the floor) is equal to that of the air. Because the rate of conductive heat transfer normally is quite low, the differential of heat loss between the substrate and air usually can be ignored. Surface temperature is also very difficult to measure in rodents; thus, internal or core temperature is substituted for this variable. For temperature differences of 20°C or less, the Stefan-Boltzmann principle of radiant heat exchange can be disregarded, and a linear relationship for radiant heat loss can be assumed. Moreover, it is also quite difficult, if not impossible, to simultaneously measure A_r, A_k, A_c, h_k, and h_c. But thermal physiologists working with rodents have found that equation (1.4) can be further simplified to

$$M = C'(T_b - T_a) + \lambda\text{EWL} \qquad (1.5)$$

where C' is whole-body thermal conductance and is approximately equal to $4\epsilon\sigma A_r T_a^3 + h_k A_k + h_c A_c$ (McNab, 1980). Clearly, thermal conductance is an extreme simplification of all the complex avenues of dry heat loss, but it is indeed a useful parameter for rodent thermophysiological studies (see Chapter 3 for further discussion). More detailed discussions of heat transfer

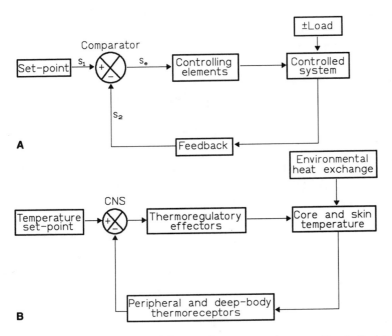

Figure 1.2. A: General diagram of a servo control system with a feedback loop. B: Basic model of thermoregulatory control system using the concept of a servo control system. Model adapted from Schmidt-Nielsen (1975b) and Stolwijk and Hardy (1974).

and its application to the thermal biology of rodents and other homeotherms are available from various sources (Monteith and Mount, 1974; Monteith and Unsworth, 1990; McArthur, 1991).

1.4.3 Control theory: the concept of the set-point

Thermoregulation in homeotherms typifies the concept of a servoloop-regulated control system (Figure 1.2A). In such a system an actuating or error signal (S_e) is generated in a comparator device by the summation of a set-point or reference signal (S_1) and a feedback signal (S_2). Any disturbance in the output of the controlled system is detected by feedback, resulting in an error signal and the activation of a corrective response.

The simple feedback system in Figure 1.2A has been quite useful for modeling the control of physiological phenomena, including the control of body temperature in rodents and other species. The thermoregulatory system can be modeled into a control loop with four major components: thermal receptors (feedback), CNS integrative and control neurons (comparator), heat-producing, -conserving, and -dissipating effectors (controller), and a passively controlled system (core and skin temperatures) (Figure 1.2B). Although the thermoregulatory characteristics of rodents differ in many respects from those of other mammals, the concept depicted in the feedback loop of Figure 1.2B is essentially universal for all homeothermic species.

The thermoregulatory system consists of two primary feedback loops: one for the regulation of heat gain/heat conservation, and one for the regulation of heat dissipation. For rodents, in particular, each of the components in Figure 1.2B can be expanded to show the basic thermoregulatory reflexes (Figure 1.3). In this model, maintenance of a constant core temperature is achieved through an elevation in metabolic heat production and a reduction in heat loss when heat storage is negative, and a reduction in metabolic heat production and an increase in heat loss when heat storage is positive. Indeed, if one records total heat loss and heat production in a rodent or other homeotherm in steady-state conditions, a clear pattern of waxing and waning of positive and negative heat storage is seen that reflects subtle changes in heat gain and loss (Figure 1.4). When the positive and negative heat-storage terms are summed over a suitable period of time, they add to zero, indicative of steady-state regulation of core temperature.

Although the set-point is clearly definable in the control system of Figure 1.2, this term has become rather enigmatic in thermal physiology. The set-point often is viewed as being analogous to the setting of the temperature of a thermostat for a mechanical temperature controller. However, several decades ago research on the thermal sensitivity of the brain stem indicated that regulation was achieved by a variable set-point whose value was dependent on ambient temperature, fever, exercise, and other factors. The debate over the existence of a neurological mechanism for a set-point continues to be a key topic in thermoregulation (for details, see Chapter 2). It is currently defined as "the value of a regulated variable which a healthy organism tends to stabilize by the processes of regulation" (IUPS, 1987). In other words, it can be inferred that the set-point for control of body temperature is maintained at this level with no apparent dysfunction.

Whether or not there is a real set-point and how it is generated in the CNS are inconsequential matters at this point. More important for this discussion is that the thermoregulatory system in rodents and other mammals respond as if there were a set-point of $37 \pm 1°C$. The concept of the set-point is immensely helpful in explaining thermoregulatory responses. There are five basic thermoregulatory states that may or may not involve a change in set-point: normothermia ($T_b = T_{set}$), regulated hyperthermia ($T_{set} > T_b$), forced hyperthermia ($T_b > T_{set}$), regulated hypothermia ($T_{set} < T_b$), and forced

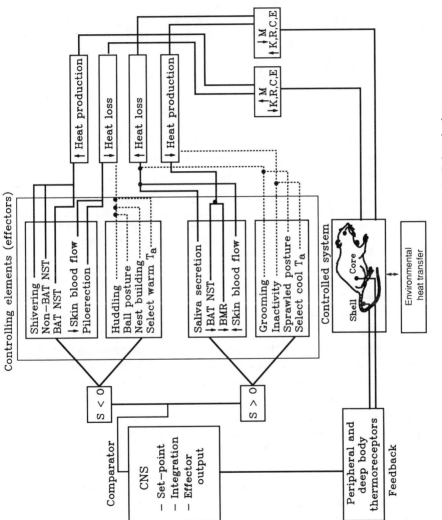

Figure 1.3. Expansion of the components in Figure 1.2 to show detailed mechanisms of thermoregulatory control in a "typical rodent."

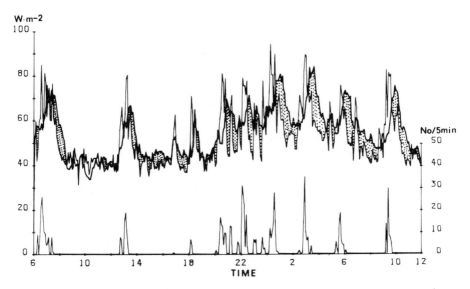

Figure 1.4. Time courses of heat production (*M*; thin line), heat loss (*H*; thick line), and feeding activity (bottom line) for a rat over a 24-hr period. Shaded areas represent negative heat storage. Note overall increases in *M* and *H* and increased feeding activity during nighttime. Light cycle: 0500 to 1900 hr. Reprinted from Sugano (1983) with permission from Pergamon Press.

hypothermia ($T_b < T_{set}$) (Figure 1.5). In this model system, during normothermia the body temperature exhibits subtle oscillations above and below the set-point temperature, and thermoregulatory tone is essentially normal. During regulated hyperthermia, an agent affects the CNS, resulting in an elevation in the set-point. In this case, the animal reacts as if it were cold, and it generates appropriate heat-gain/conserving effectors to raise its core temperature to equal the new set-point. Fever is a classic example of an elevation in the set-point. During forced hyperthermia, the body temperature is forced above the set-point, as would occur during severe heat stress or administration of a drug that stimulates metabolism without affecting the set-point. In this case, heat-dissipating effectors are activated. During regulated hypothermia, as is thought to occur during the onset of hibernation, the set-point is reduced below the core temperature, and the animal acts as if it were hot and activates heat-dissipating responses to lower its body temperature. This same effect occurs when a subject with a fever is treated with an antipyretic to lower the set-point: the subject begins to sweat and reacts as if he were hot. During forced hypothermia, the body temperature is reduced below the set-

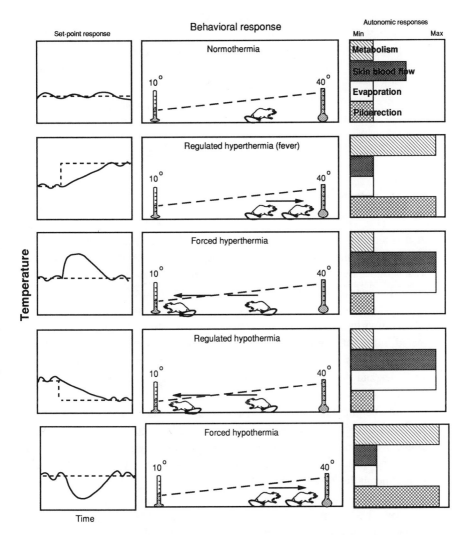

Figure 1.5. Summary of behavioral and autonomic thermoregulatory responses of a rodent when subjected to manipulations of body and set-point temperatures: normothermia, regulated hyperthermia, forced hyperthermia, regulated hypothermia, and forced hypothermia. Modified from Gordon (1983c).

point, as might occur during immersion in icy water or administration of a drug that inhibits metabolism but does not affect the CNS.

Measuring behavioral thermoregulatory responses can be quite helpful in elucidating whether or not the set-point has changed. Such measurements have been widely used in studying the neuropharmacology of temperature regulation (Satinoff, 1978; Gordon, 1983c; Marques et al., 1984; Gordon et al., 1988). For example, if after treatment with a chemical purported to lower the set-point a rat exhibits vasodilation of the tail, a reduction in metabolic rate, and hypothermia, one might conclude that this chemical is indeed lowering the set-point. However, these effects could be mediated systemically without affecting the CNS. If the thermoregulatory behavior is also measured and it is found that the animal prefers a cooler environment and also becomes hypothermic, then the conclusion of a reduction in the set-point is more tenable.

1.5. Do rodents provide a suitable model for human thermoregulation?

A fundamental premise in biomedical research is that data from studies of experimental animals can, to varying degrees, be extrapolated to humans. When there are significant dissimilarities between the experimental model and the human, then appropriate scaling factors or other manipulations must be incorporated to normalize the species-to-species variations.

An important question then arises: Are the thermoregulatory data collected in the rat or other experimental rodents relevant to human responses? To address this issue, I have listed some of the pertinent thermoregulatory variables and their approximate values in a 0.25-kg rat and an 80-kg human (Table 1.1). Considering the 320-fold difference in their body masses, one can only be impressed with the similarities in the magnitudes of many of the variables in the two species, including core body temperature, skin temperature, preferred temperature, lower critical ambient temperature, and upper lethal body temperature. On the other hand, there are enormous differences in the metabolic parameters, including basal metabolic rate (both total heat loss and heat loss per unit body mass), peak metabolism, and evaporative heat loss. The metabolic dissimilarities are attributable primarily to the differences in body mass and surface areas and are explained in more detail in Chapter 3. In general, the rat (like other small rodents) has a relatively large surface area : body-mass ratio and must maintain a high rate of heat production to keep its core and skin temperatures equivalent to those of the human (or most other large mammals). Thus, the rat and the human have similar core and skin temperatures but have totally different thermoregulatory "strategies." That is, for the rat to maintain its core and skin temperatures equivalent to those of a human, its heat production per gram of body weight must be approximately five times that of the human.

The data in Table 1.1 clearly present an oversimplification, and they are

Table 1.1. *Comparison of some basic thermal physiological variables in a "typical" laboratory rat and human[a]*

Variable	Rat	Human	Human/Rat
Body weight (kg)	0.25	80	320
Core temperature (°C)	37	37	1
Skin temperature (°C)	~30	33	1.1
Selected T_a (°C)	28	28	1
Lower critical T_a (°C)	28	28	1
Lethal core temperature			
Upper (°C)	44	43	0.97
Lower (°C)	15	26.8	1.8
Surface area : body mass ($m^2\ kg^{-1}$)	0.13	0.025	0.2
Metabolism			
(W)	1.3	101	77.7
($W\ kg^{-1}$)	5.3	1.3	0.24
($W\ m^{-2}$)	38.4	50.4	1.3
Maximum metabolism ($W\ kg^{-1}$)	33.5	11.3	0.33
Maximum evaporation ($W\ kg^{-1}$)	8.8	17.5	1.9

Note: [a] Physical variables such as surface area were calculated from allometric scaling equations. The physiological variables given are derived from various sources in the literature.

used here only to illustrate basic differences between the rat and the human. These data demonstrate some of the difficulties that may be encountered in attempting to extrapolate data from a laboratory rodent to a human. For example, treatment with a drug or chemical agent that affects metabolism will have, at least in the short term, a more marked effect on the rat's thermoregulatory output. Also, to maintain a constant body temperature under a variety of environmental challenges (cold stress, food deprivation, exercise, hypoxia, etc.), the rat's thermoregulatory effectors will exhibit relatively greater responses than will those of the human. This should consequently affect the activity and regulation of other physiological systems more strongly in the rat (caloric requirements, cardiovascular system, water balance, etc.). Thus, although we rely heavily on the rat and other laboratory rodents to provide a first-line estimate for potential effects in humans, we must continually be aware of how these thermophysiological differences can affect the interpretation and extrapolation of the data.

2

Neurology of temperature regulation

Understanding the neurological processes of temperature regulation has been a challenge to researchers in a variety of disciplines over the past several decades. Four fundamental methods have been used to study the neural control of body temperature in rodents and other species: (1) thermal stimulation of CNS sites with stereotaxically implanted thermodes; (2) neurophysiological recordings of the firing rates of single neurons in thermal afferent and integrative pathways; (3) neuropharmacological stimulation of thermoregulatory pathways and measurement of neurohumoral substances in the CNS during thermal stimulation; (4) lesions and/or ablations of CNS sites. These and other methods, when applied to normal and febrile subjects, have led to the development of various regulatory models of temperature regulation. This chapter is an attempt to discuss the key aspects of what has become an enormous subdiscipline of thermal physiology.

2.1. Temperature sensitivity of the CNS

Since the 1930s it has been established that sites in the hypothalamus and preoptic area elicit thermoregulatory motor responses when thermally stimulated; for historical reviews, see Lomax (1979) and Schönbaum and Lomax (1990). There has been a major research effort to discern the comparative physiological aspects of the thermal sensitivity of the CNS. The preoptic area/anterior hypothalamus (POAH) and other sites in the CNS in rodents and other mammals are extremely sensitive to artificial displacements above or below the normal brain temperature of ~37°C (Table 2.1). For example, localized warming of the POAH elicits heat-dissipating (i.e., thermolytic) responses, including peripheral vasodilation, increased evaporative water loss, reduced metabolism, and a preference for cooler ambient temperatures. When these effectors are stimulated, heat loss will exceed heat gain, resulting in

Table 2.1. *Survey of studies demonstrating temperature sensitivity in the CNS in laboratory rodents*

Species	CNS site of stimulation	Thermoregulatory responses Cooling	Warming	References
Hamster	spinal cord	↑ M		Wünnenberg (1983)
Hamster	POAH	↑ M	↓ SH	Wünnenberg (1983)
Rat	spinal cord	↑ M, VC, ↑ T_b	↓ M, VD, ↓ T_b	Banet and Hensel (1976); Banet et al. (1978)
Rat	spinal cord	↓ R, TC	VD, ↑ R, BC, ↓ T_b	Lin and Chai (1974)
Rat	medulla	VC, ↓ R, TC, ↑ T_b	VD, ↑ R, BC, ↓ T_b	Lin and Chai (1974)
Rat	medulla	↑ T_b, ↑ ST_a	↓ T_b, ↓ ST_a	Lipton (1973)
Rat	POAH	↑ T_b, ↑ ST_a	↓ T_b, ↓ ST_a	Lipton (1973)
Rat	POAH	↑ M, VC		Banet et al. (1978)
Rat	POAH	SH, ↑ ST_a, ↑ T_b		Satinoff (1964)
Rat	POAH	↓ FC, ↑ T_b	↑ FC, ↓ T_b	Spector et al. (1968)
Rat	POAH		GM, BE	Roberts and Mooney (1974)
Guinea pig	POAH		↓ M	Brück and Zeisberger (1978)

Note: Key: M, metabolism; TC, tachycardia; BC, bradycardia; VC, peripheral vasoconstriction; VD, peripheral vasodilation; ST_a, selected T_a; GM, saliva grooming; BE, body extension; T_b, core body temperature; SH, shivering; FC, food consumption; R, respiration rate.

negative heat storage and a decrease in body temperature (Figure 2.1). On the other hand, cooling the POAH stimulates heat-conserving and heat-producing motor responses, including peripheral vasoconstriction, shivering and nonshivering thermogenesis, and a preference for warmer temperatures. Stimulating these effectors causes heat gain to exceed heat loss, resulting in positive heat storage and an elevation in body temperature.

The thermal-sensitive nature of the POAH has been well characterized in species such as rabbit, cat, dog, and goat (Hammel, 1968; Hensel, 1973, 1981; Cabanac, 1975). Unfortunately, relatively few studies have been carried out in commonly used laboratory rodents, thus making a comparative analysis in these species difficult. In spite of the fact that there have been relatively few studies, many of the findings collected from other mammalian species can be applied to analyze the thermal characteristics of the POAH and other sites in the rodent CNS.

2.1.1. Proportional control in the POAH

When thermodes are implanted into the CNS via stereotaxic surgery and are continuously perfused with water of varying temperatures, the temperature of the POAH or other sites in the CNS can be "clamped" while one or more effector responses are measured in the unrestrained animal. Additional probes

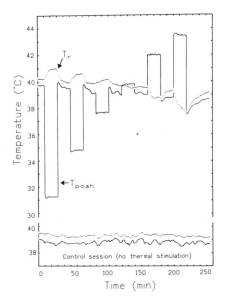

Figure 2.1. Example of the effect of artificial displacement in POAH temperature (T_{poah}) on the rectal temperature (T_r) of a rat. Note that the change in rectal temperature is opposite to the change in POAH temperature. Modified from Corbit (1970).

can be implanted to monitor the temperature and/or to inject neurochemicals within a given site in the CNS. Rodents easily tolerate these procedures and usually recover their normal thermoregulatory reflexes within 1 week after surgery. A fundamental neural model for thermoregulatory control in mammals was developed in the 1960s as a result of extensive research on the thermal sensitivity of the POAH; for reviews, see Hammel (1968), Heath et al. (1972), Hensel (1973), and Simon (1981). Basically, the thermal sensitivity of the POAH can be modeled as a proportional controller, with input–output responses described by a simple linear equation:

$$R - R_0 = \alpha(T_h - T_{set}) \tag{2.1}$$

where R_0 is the basal motor response, R is the motor response when T_h is greater than or less than T_{set}, α is the POAH proportionality constant for a given motor response (negative for heat production/conservation; positive for heat dissipation), T_h is the POAH or hypothalamic temperature, and T_{set} is the set-point or threshold temperature. This equation can predict either heat-producing/conserving motor responses or heat-dissipating responses, depending on the direction of change in T_h and the motor response measured. The sign of the difference between T_h and T_{set} determines whether the motor response is for heat production/conservation (negative) or heat dissipation (positive). Values for α and T_{set} are derived for individual thermoregulatory motor responses such as metabolism, evaporation, peripheral vasomotor tone, and behavioral thermoregulatory responses. It can be seen that the magnitude of a given thermoregulatory output (i.e., R-R_0) at a constant T_h can be influenced by both α and T_{set}. It is important to note that equation (2.1) was developed from studies where the POAH temperature was artificially maintained at levels that usually were far above or below those normally experienced by most mammals, even under wide variations in ambient temperature (see Chapter 5). Yet it is the artificial displacement in brain temperature that has led to the characterization of the set-point. Equation (2.1) can also be used to model the characteristics of other temperature-sensitive sites in the CNS, such as the medulla and spinal cord.

The term "set-point" is somewhat of an enigma and deserves special attention. To most people, a set-point temperature would imply a fixed value, as one would find on the setting of a mechanical or electronic thermostat used for controlling room temperature. That is, the set-point would be a threshold temperature, and when the room temperature deviated from the set-point, a corrective motor response would be elicited (e.g., activating a furnace or air conditioner). However, the set-point temperature in equation (2.1) is not a fixed term as in a thermostat. It is considered to be variable, and its magnitude is affected by a combination of the intrinsic POAH temperature and a term dependent on the activity of cutaneous thermal receptors, exercise, fever, sleep/wake status, and presumably any other internal or external factors that affect thermoregulatory output (Hammel, 1968; Hensel, 1981). The operation of a variable set-point can be demonstrated by comparing the metabolic rates of a mammal at thermoneutrality ($T_a = 30°C$) and during cold exposure ($T_a = 10°C$) while assuming that the proportionality constant is fixed:

If $T_a = 30°C$, $T_{set} = 37°C$, $T_h = 37°C$, $\alpha = -1.0$ W kg^{-1} °C^{-1},
  and $R_0 = 5.0$ W kg^{-1}, then $R = 5.0$ W kg^{-1}.

If $T_a = 10°C$, $T_{set} = 40°C$, $T_h = 37°C$, $\alpha = -1.0$ W kg^{-1} °C^{-1},
  and $R_0 = 5.0$ W kg^{-1}, then $R = 8.0$ W kg^{-1}.

In this case, because of increased activity of cold thermoreceptors, the POAH responds as if T_{set} were increased by 3.0°C. Thus, even though T_h is unchanged during cold exposure, the metabolic rate is nonetheless increased because of the elevation in T_{set}. Most researchers eschew the concept

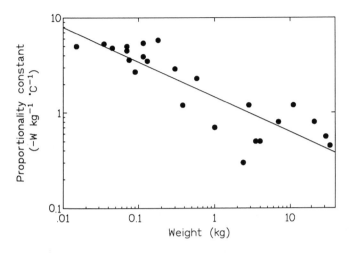

Figure 2.2. Relationship between body mass and proportionality constant for metabolic heat production (α_{hp}) measured in a variety of rodents and other mammals. $Y = 1.47X^{-.366}$. Data from Heller (1978).

of equation (2.1) and conceptualize the set-point as a fixed value, analogous to a thermostat. In that view, the set-point temperature is considered to be equal to the core or brain temperature to which an organism stabilizes by its regulatory processes (i.e., $37 \pm 1°C$ for most mammals). That is an established definition that can be applied in most thermoregulatory studies. Yet the alternative view of T_{set} in equation (2.1) as a dynamic variable is also an essential concept for modeling the neural processes of thermoregulatory control.

The principle of equation (2.1) has been used primarily to study the thermal sensitivity of the rodent CNS to cooling. Metabolic heat production can be easily measured in unrestrained rodents while the POAH temperature is clamped at various levels. The proportionality constant for eliciting an increase in heat production (i.e., a_{hp}) has been determined for a variety of rodents and other mammals subjected to POAH cooling (Heath et al., 1971; Heller, 1978). Heller (1978) found an inverse linear relationship between log body mass and log α_{hp} measured during POAH cooling in a variety of mammals, including rodent and nonrodent species (Figure 2.2). It is clear from this relationship that the POAH sensitivity to cooling is inversely dependent on body mass. This implies that smaller species should have a greater central thermoregulatory drive, whereas large species should rely more on peripheral or cutaneous thermal stimuli to control their thermoregulatory responses.

Table 2.2. *Comparisons of TBTS and local thermosensitivity in the POAH for several laboratory rodents*

Species	TBTS $(W\ kg^{-1}\ {}^{\circ}C^{-1})$	α_{hp} $(W\ kg^{-1}\ {}^{\circ}C^{-1})$
Mouse	−9.3	?
Hamster	−8.7	?
Rat	−4.7	−0.4
Guinea pig	−5.7	−0.6

Source: Modified from Mercer and Simon (1984).

It will be noted that the format of equation (2.1) is similar to that used for modeling the metabolic responses of homeotherms at ambient temperatures below thermoneutrality [i.e., $M = C'(T_b - T_a)$; see Chapter 3, equation (3.6)]. That is, the metabolic responses of most mammals to ambient cooling are proportional to two temperature differentials: (1) core and ambient and (2) hypothalamic and T_{set}. Moreover, as with the inverse dependence of α_{hp} on body size discussed earlier, C' also increases with decreasing body mass. Heller (1978) proposed that the increased POAH thermal sensitivity in smaller species is adaptive to their greater thermal conductance and hence their increased tendency to lose heat during cold stress. A confound of this proposed mechanism is that the temperature of the rodent brain is quite stable over a wide range of ambient temperatures (see Chapter 5). Thus, even with a high α_{hp}, one must question how the metabolic rate is controlled in spite of a normal brain temperature. It is also not clear how T_{set} may vary as a function of body mass. Although proportional control explains metabolic responses under experimental conditions, it is not clear that changes in brain temperature under natural conditions have any role in governing thermoregulatory responses in rodents.

Thermal stimulation of local sites in the CNS is indeed a useful way of understanding thermoregulatory control. However, thermal receptors are found throughout the body, and the stimulation of just one site does not necessarily provide a measure of the animal's total-body thermosensitivity (TBTS). Measuring the TBTS can be done by clamping the entire body temperature above or below normal using intravascular or intraabdominal heat exchangers (Mercer and Simon, 1984). TBTS values are markedly higher in the mouse and hamster than in the rat and guinea pig (Table 2.2). TBTS values for the rat and guinea pig are approximately 10 times greater than the value of α_{hp} for these species determined by hypothalamic cooling. For the guinea pig, the thermosensitive responses of the spinal cord and POAH contribute approximately 20% of the TBTS during whole-body cooling (Mercer and Simon, 1984). The measure of TBTS illustrates the importance of thermal afferents other than the POAH for driving metabolic thermogenesis in these rodent species. Moreover, the fact that smaller species like the mouse and hamster have higher TBTS values demonstrates (as with the POAH thermal sensitivity discussed earlier) increased sensitivity of metabolic thermogenesis in species of small body mass.

2.2. Neurophysiological studies

The laboratory rat and guinea pig have been used extensively to study many neurophysiological facets of temperature regulation. Research in rodents has centered on understanding (1) the encoding of thermal stimuli into neural signals by cutaneous thermoreceptors, (2) the transfer of thermoafferent information through the CNS, and (3) the integration and temperature-sensing characteristics of neurons in the POAH and other sites in the CNS. The generation of effector signals by the CNS for control of heat loss and heat production is not nearly as well understood in thermoregulation as in other systems. Specific aspects of central neural control of thermoregulatory effectors are covered in the next chapter.

2.2.1. Thermoafferent system

Like other mammals, rodents have distinct warm and cold receptors that display the characteristic dynamic and static firing-rate responses when stimulated with abrupt and prolonged changes in temperature (Figure 2.3). Generally, the static discharge activity peaks at skin temperatures of 25–27°C for cold receptors and 40–45°C for warm receptors (Spray, 1986). The scrotum of the rat (and presumably other rodents) is highly responsive to temperature and has been used extensively to study the characteristics of thermoreceptors and the properties of the thermoafferent system at the spinal and supraspinal levels. Isolated heating or cooling of the rat scrotum stimulates heat-dissipating or -conserving responses, respectively (see Chapter 6). The firing rates of specific warm and cold thermoreceptors in the rat scrotum display both a dynamic overshoot (i.e., phasic response) to a change in temperature and a static or nonadapting response (i.e., tonic response) (Figure 2.3). The maximum activity of the tonic response occurs at skin temperatures of 31°C for cold receptors and 42°C for warm receptors (Hellon, Hensel, and Schäfer, 1975). Specific cold receptors also begin to fire at temperatures

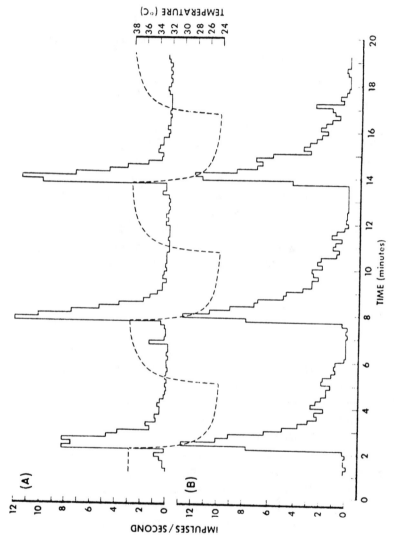

Figure 2.3. Example of the dynamic overshoot and static response of a cutaneous cold-responding mechanoreceptor (A) and cold thermoreceptor (B) in the scrotum of a rat. Reprinted from Pierau et al. (1975) with permission from The American Physiological Society.

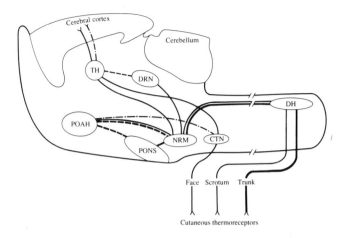

Figure 2.4. Sagittal view of the rat brain showing the three major thermoafferent pathways for the transfer of cutaneous thermal information from the face, trunk (thick lines), and scrotum (thin lines) to the CNS. TH, thalamus; DRN, dorsal raphe nuclei; NRM, nucleus raphe magnus; CTN, cutaneous trigeminal nuclei; DH, dorsal horn of spinal cord. Solid lines, established connections; dash lines, presumed connections (trunk and scrotum); dot-and-dash lines, presumed connections (face). Modified from Hellon (1983).

above 45°C, a paradoxical phenomenon noted in other species (Spray, 1986). In addition to specific thermoreceptors, there are mechanosensitive neurons in the rat scrotum that also respond to thermal stimulation, and display static or dynamic responses to skin cooling (Pierau, Torrey, and Carpenter, 1975).

Using the anesthetized rat as the principal experimental model, R. F. Hellon and others have shown the operation of three pathways originating from the face, trunk, and scrotum for the passage of cutaneous thermal stimuli through the CNS (Figure 2.4). Thermal information from the scrotum and most of the skin of the trunk is conveyed through the dorsal horn of the spinal cord, through the nucleus raphe magnus, and then on to the POAH, thalamus, and other CNS locations. Thermal stimuli from the face are first processed in the trigeminal nucleus and then are thought to be conveyed to the POAH and thalamus for further processing. Single-neuron recordings in the CNS have revealed significant modifications of the neural activity originally encoded by the cutaneous thermoreceptors. The characteristic dynamic response of cutaneous thermoreceptors is markedly dampened by the time the information reaches the spinal cord and CNS nuclei. Generally, dynamic responses to rapid changes in skin temperature are not easily detected in the CNS (Hellon and Misra, 1973a,b; Nakayama et al., 1983). In the CNS,

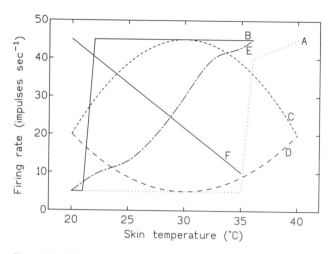

Figure 2.5. Principal neuronal responses of thermoafferent neurons to changes in skin temperature at the spinal and supraspinal levels: A, switching response; B, inverse switching response; C, bell-shaped response; D, inverse bell-shaped response; E, sigmoid (warm-sensitive) response; F, linear response (cold-sensitive).

thermoafferent information is encoded with a variety of functions: sigmoid, linear, bell-shaped, and switching or on–off responses (Figure 2.5). The functions with relatively small changes in firing rate per 1°C change in skin temperature (i.e., bell-shaped, linear, and sigmoid) have been recorded at various levels of the CNS, including neurons in the dorsal horn of the spinal cord (Hellon and Misra, 1973a; Pierau et al., 1984), trigeminal nucleus (Dawson et al., 1982), raphe nuclei (Dickenson, 1977; Hellon and Taylor, 1982), pons (Hinckel and Schröder-Rosenstock, 1981), thalamus (Hellon and Misra, 1973b; Schingnitz and Werner, 1979), and hypothalamus (Kanosue et al., 1984). Because of the relatively small changes in the activity curves for these neurons in response to temperature changes (i.e., impulses per second per 1°C), it is thought that they probably are important for measurements of graded changes in skin temperature. Such a form of detection of temperature would be essential for the control of thermoregulatory effectors that exhibit proportional responses to changes in skin or ambient temperature, such as metabolic rate and skin blood flow (Brück and Hinckel, 1980; Hinckel and Schröder-Rosenstock, 1981; Dawson et al., 1982).

The operation of switching neurons is a relatively recent discovery in the thermoafferent system. Switching neurons have a characteristic step function in their firing activity. They exhibit little or no activity at subthreshold temperatures, but exhibit an abrupt elevation in firing rate as temperature is increased by as little as 0.5°C above that threshold (Nakayama, Ishikawa, and Tsuratani, 1979; Schingnitz and Werner, 1979; Hellon and Taylor, 1982; Taylor, 1982; Pierau et al., 1984). There are also inverse switching neurons that have a relatively high level of activity and then show abrupt inhibition as skin temperature is changed at a critical level. The role of switching neurons in temperature regulation remains unclear. Their temperature–activity profile suggests an alarm or trigger signal for heat dissipation, specifically for the protection of the testes from overheating (Werner, Schingnitz, and Mathei, 1986). Switching neurons are predominantly activated by scrotal thermal stimulation. Thus, it is possible that switching neurons are involved in a reflex for the protection of spermatogenesis against thermal stress. Interestingly, warming the skin over the genital area of the female rat does not generate switching responses in the thalamus; however, such responses are elicited in the female rat during thermal stimulation of the abdominal skin (Schingnitz and Werner, 1986). Moreover, switching responses also occur in the guinea pig thalamus during abdominal stimulation, but not scrotal stimulation (Schingnitz and Werner, 1986). Thus, it appears that the switching-response neurons are common in the thermal afferent system in rodents, but there are species- and gender-dependent differences in organization of the thermoafferent pathways.

Interestingly, its been shown that sectioning the cutaneous sensory fibers innervating the trunk and face has a surprisingly meager effect on the ability of the rat to thermoregulate in the cold (Heath, 1985). Following surgical denervation in which the majority of cutaneous thermoreceptors are rendered nonfunctional, the major thermoregulatory deficits are increased thermal conductance and an elevated metabolic rate during ambient cooling and a 2.1°C increase in the lower critical ambient temperature (see Chapter 3 for terminology). The ability to control body temperature in spite of such drastic impairment in the thermoafferent system illustrates the resilience of the thermoregulatory system.

It would appear that temperature-sensitive neurons in the POAH possess a degree of laterality for the activation of some thermoregulatory effectors. For example, when the right or left side of the POAH of the anesthetized rat is heated, the resultant vasodilation and salivation are more pronounced on the ipsilateral side, whereas hindlimb shivering is inhibited equally on the ipsilateral and contralateral sides (Kanosue et al., 1991). Unilateral skin heating of the abdomen results in uniform, bilateral vasodilation, salivation, and suppression of shivering. Hence, thermoafferent information from the skin probably converges before reaching the POAH. Thermal-sensitive neurons within the POAH apparently have more direct connections for the activation of salivation and peripheral vasomotor tone, whereas the efferent pathway of shivering lacks the lateral organization.

2.2.2. POAH neuronal properties

Whole-animal studies.   The literature abounds with studies on neurophysiological and thermal-sensitive characteristics of the POAH; for reviews, see Hammel (1968), Hensel (1973), Cabanac (1975), Gordon and Heath (1986), and Boulant and Dean (1986). The POAH has been characterized neurophysiologically (1) in anesthetized rodents by recording the activity of single neurons in vivo under direct thermal stimulation via stereotaxically implanted thermodes and (2) in vitro, in the absence of anesthetics, by recording neuron activity in brain-slice preparations maintained in temperature-controlled media (Table 2.3). In general, the POAH of the anesthetized rat consists of 30% warm-sensitive neurons, 10% cold-sensitive neurons, and 60% insensitive neurons. Although the numbers of studies in other rodents have been relatively meager, the available data suggest that the proportions of temperature-sensitive neurons in the guinea pig, hamster, and mouse are similar to those in the rat.

Neurons with varying degrees of thermal sensitivity have been isolated in most parts of the CNS in rodents. However, that POAH neurons exhibit increases in firing rates during heating or cooling does not necessarily warrant that the cells are involved in regulatory processes (Hensel, 1973). That is, a change in brain temperature should have direct effects on the activities of most neurons in the CNS. To circumvent this problem, various criteria have been developed to distinguish between temperature-sensitive and -insensitive neurons (e.g., Boulant and Dean, 1986; Boulant, Curras, and Dean, 1989). Generally, neurons are classified as warm-sensitive if either the Q_{10} (see Section 3.6) is greater than 2.0 or the regression coefficient (i.e., Δ firing rate/Δ temperature) is equal to or greater than 0.8 impulse per second per degree Celsius; neurons are classified as cold-sensitive if $Q_{10} \leq 0.5$ or the regression coefficient is equal to or less than -0.6 impulse per second per degree Celsius. Neurons with coefficients between these values are classified as temperature-insensitive. Neurons meeting these criteria could nonetheless be simply temperature-sensitive but play no role in thermoregulatory control. On the other hand, many temperature-sensitive POAH neurons are also responsive to peripheral thermal stimulation of the scrotum, tail, and face (Figure 2.6). The firing rate of POAH neurons can also be modulated by thermal stimulation of other sites in the CNS, such as the midbrain, medulla, and spinal cord (Gordon and Heath, 1986). The presence of impinging warm and cold pe-

Table 2.3. *Survey of recordings of temperature-sensitive and -insensitive neurons in various locations in the CNS in laboratory rodents*

Species	CNS location	Neuron distribution (%)			References
		Warm	Cold	Insensitive	
Mouse[a]	PO, AH (1)	28	4	68	Boulant and Dean (1986)
Hamster[b]	PO (1)	24	7	69	Wünnenberg et al. (1976)
Rat[b]	PO, AH (11)	35	13	52	Boulant and Dean (1986)
Rat[b]	PO (3)	37	13	51	Boulant and Dean (1986)
Rat[c]	POAH (5)	34	8	58	Boulant and Dean (1986)
Rat[b]	MB (1)	58	0	42	Boulant and Dean (1986)
Guinea pig[c]	POAH (1)	43	13	44	Boulant and Dean (1986)
Guinea pig[c]	PO (1)	43	12	45	Boulant and Dean 1986
Guinea pig[b]	PO (1)	20	3	67	Wünnenberg et al. (1976)

Key: PO, preoptic area; AH, anterior hypothalamus; MB, midbrain. Some of the neuron distributions represent averages of more than one study, as presented Boulant and Dean (1986). Numbers in parentheses represent numbers of studies averaged.
[a]Neurons recorded in tissue culture.
[b]Neurons recorded in anesthetized animals.
[c]Neurons recorded in brain slices.

ripheral inputs onto temperature-sensitive neurons in the CNS certainly increases the likelihood that these neurons are involved in thermoregulatory control. Moreover, temperature-insensitive neurons in the CNS can also receive input from thermoafferent pathways and can play a role in controlling thermoregulatory responses.

Brain slice. The preoptic area and hypothalamus are involved in the regulation of many homeostatic processes in addition to temperature regulation, including sexual behavior, reproduction, water balance, food intake, and others. These homeostatic processes do not operate independently. That is, behavioral and physiological studies have shown that the control of one variable can be influenced by the activity of another variable. Thus, neurons in the POAH are likely to be focal points for the integration of multiple autonomic processes. Studies by J. A. Boulant and others using brain-slice preparations have clearly demonstrated the sensitivity of temperature-sensitive neurons to

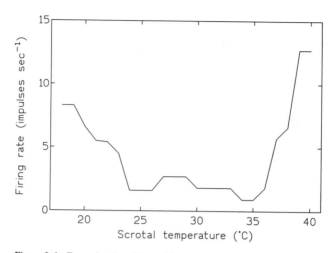

Figure 2.6. Example of a cold-sensitive POAH neuron that exhibits an inverted bell-shaped firing response to changes in temperature of the scrotal skin. Data from Nakayama et al. (1979).

nonthermal stimuli. Indeed, the concept of "functional specificity" (i.e., the theory that each neuron responds to single factors) appears to be rarely applicable in the POAH (Boulant and Silva, 1989). Individual temperature-sensitive and -insensitive POAH neurons respond to various endogenous factors such as hypoglycemia, hypoosmolality and hyperosmolality, and reproductive steroids (Figure 2.7). In one instance, the interactions between luteinizing-hormone-releasing hormone and the activities of warm- and cold-sensitive neurons can explain why reproductive activity is suppressed during exposure to warm temperatures (see Chapter 6). The advantage of the brain-slice preparation is that neurons can be studied in the absence of anesthetics and can be easily stimulated with a variety of neurotransmitters, hormones, and other agents. A drawback of this technique is its obvious absence of afferent stimulation, which undoubtedly alters the natural characteristics of many POAH neurons. In spite of these limitations, the brain-slice technique promises new breakthroughs in the field of neurophysiological control of body temperature; for reviews, see Boulant et al. (1989) and Hori (1991).

Hibernation. It is interesting to compare the thermal-sensitive characteristics of POAH neurons form the golden hamster, a hibernator, and those of nonhibernating species, such as the rat and guinea pig (Table 2.4). A large percentage of neurons in the hamster continue to fire at brain temperatures below 20°C. Neurons from the rat also fire at cooler temperatures, but the

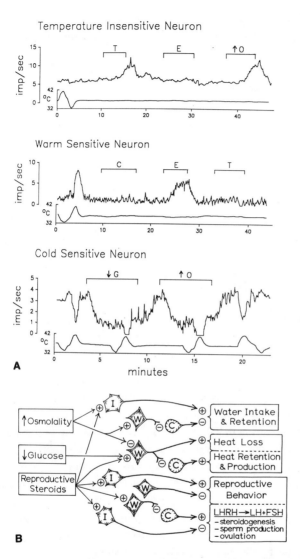

Figure 2.7. A: Responses of temperature-insensitive, warm-sensitive, and cold-sensitive POAH neurons recorded from a rat brain slice during steroidal, hyperosmotic, and low-glucose stimulation. Stimuli include 30 pg/ml testosterone (T), 30 pg/ml estradiol (E), 0.002% ethanol control (C), 309 mosmol hyperosmotic (O), and 1 mM glucose (G, low glucose). B: Neuronal model of thermoregulatory control in rat based primarily on in vitro brain-slice recording data: plus sign indicates excitation; minus indicates inhibition; LHRH, luteinizing-hormone-releasing hormone; LH, luteinizing hormone; FSH, follicle-stimulating hormone. Modified from Boulant and Silva (1989) with permission from The American Physiological Society.

Table 2.4. *Approximate percentages of warm-sensitive neurons that remain active at various brain temperatures in a hibernator (golden hamster) and nonhibernators (rat and guinea pig)*

Species	T_{POAH} (°C)	Neurons remaining active (%)
Hamster	35–40	100
	30–35	100
	25–30	100
	20–25	100
	15–20	100
	10–15	~33.3
Rat	35–40	100
	30–35	92.3
	25–30	92.3
	20–25	69.2
	15–20	53.8
	10–15	23.1
Guinea pig	35–40	100
	30–35	100
	25–30	50
	<25	0

Source: Data from Wünnenberg, Kuhnen, and Laschefski-Sievers (1986).

percentage of active neurons drops markedly at temperatures below 20°C. The guinea pig appears to have the least adaptive response to cooling, with no active neurons below 25°C. In this regard, others have found a variety of adaptations to hypothermia in the hamster's peripheral and central nervous systems. For example, conduction of action potentials in the tibial nerve ceases at 9.0°C in the rat, whereas in the hamster conduction ceases at 3.4°C (Chatfield et al., 1948). During deep hypothermia in the rat the integrity of the blood-brain barrier (BBB), as measured by permeability to compounds such as rubidium and albumin, is compromised as brain temperature decreases below ~19°C (Wells, 1972). However, the permeability of the hamster BBB remains relatively intact at brain temperatures as low as 13°C. It is evident that the nervous system in hibernators such as the hamster has evolved to function at extremely low temperatures and thereby assure thermoregulatory control during entrance into and recovery from hibernation.

2.3. Neuropharmacological agents

Feldberg and Myers (1964) first demonstrated specific thermoregulatory responses in the cat when adrenergic and serotonergic neurotransmitters were administered directly into the CNS. Their work undoubtedly provided the impetus for much of the research on neurochemical aspects of temperature regulation. As witnessed by the data tables compiled by W. Clark and colleagues (Clark and Lipton, 1983; 1985), neuropharmacological research over the past 30 years has generated countless papers on the thermoregulatory responses of rodents and other species given central and/or peripheral administrations of naturally occurring neurotransmitters, neuromodulators, pharmacological agonists and antagonists, neuropeptides, and other agents. A comprehensive review of this field would exceed the intended scope of coverage of this book. The reader is referred to specific reviews for detailed coverage of specific neuropharmacological agents, including acetylcholine (Myers, 1987; Myers and Lee, 1989), norepinephrine and serotonin (Myers, 1980; Myers and Lee, 1989), and CNS peptides (Clark and Lipton, 1983; Adler et al., 1988; Janský, 1990), and the role of cations in thermoregulation (Myers, 1982).

In spite of the tremendous research effort in this field, one is dismayed by the number of contradictory results reported from the testing of identical neuropharmacological substances in a given species (Table 2.5). Much of the variation may be attributable to differences in dosage, route of injection, microinjection technique in the CNS, animal handling (i.e., stress), environmental temperature, and other factors (Myers, 1980). Comparisons between species are hampered by the small sizes of the rodent brains. It is difficult to administer a given dose of a chemical into a constant volume of brain tissue in species with such small brains. Also, differences in thermoregulatory sensitivity as a function of ambient temperature have compounded the difficulties in making interspecies comparisons.

Recent studies by C. M. Blatteis may force researchers to reevaluate the purported thermoregulatory effects of various neurotransmitters and other pharmacological agents. Using a microdialysis technique to administer norepinephrine (NE) into the POAH of the guinea pig, Quan and Blatteis (1989) observed a clear reduction in body temperature; however, administration of NE into the same area using conventional microinjection techniques led to hyperthermia (Figure 2.8). During the typical microinjection of a chemical into the CNS, an injector cannula is extended a few millimeters beyond the end of the guide cannula, causing some physical damage to the brain tissue. It was found that this damage led to stimulation of prostaglandin E_2 (PGE_2) production in the POAH, resulting in a febrile response. Thus, the net thermoregulatory response to microinjection of NE is a combination of the hypothermic effects of NE and the hyperthermic effects of PGE_2. With microdialysis, there is no physical damage to neural tissue, and the true effects of NE on thermoregulation can be discerned. Clearly, the possibility of

Table 2.5. *Effects on body temperature of various neurotransmitters, peptides, and other neuromodulatory agents administered into the CNS in laboratory rodents*

Species	Injection site[a]	Temperature response	References
Acetylcholine			
Hamster	IVT	decrease	Janský (1978)
Hamster	POAH	increase	Simpson and Resch (1985)
Rat	IVT	increase	Myers and Yaksh (1968)
Rat	IVT	decrease	Lin et al. (1980)
Rat	POAH	decrease	Beckman and Carlisle (1969)
Rat	POAH	decrease	Crawshaw (1973)
Dopamine			
Mouse	IVT	decrease	Brittain and Handley (1967)
Rat	IVT	decrease	Hansen and Wishaw (1973)
Rat	IVT	increase	Myers and Yaksh (1968)
Norepinephrine			
Mouse	IVT	decrease	Brittain and Handley (1967)
Hamster	AH	decrease	Reigle and Wolfe (1974)
Hamster	IVT	decrease	Janský (1978)
Rat	IVT	increase	Myers and Yaksh (1968)
Rat	POAH	decrease	Avery (1972)
Rat	AH	decrease	Christman & Gisolfi (1980)
Rat	IVT	decrease	Cantor and Satinoff (1976)
Guinea pig	AH	increase	Zeisberger and Brück (1971)
Guinea pig	IVT	increase	Kandasamy and Williams (1983b)
5-hydroxytryptamine			
Mouse	IVT	decrease	Brittain and Handley (1967)
Hamster	H	no effect	Reigle and Wolfe (1974)
Hamster	IVT	decrease	Janský (1978)
Rat	IVT	increase	Crawshaw (1973)
Rat	IVT	decrease	Myers and Yaksh (1968)
Adrenocorticotropin (ACTH)			
Guinea pig	IVT	decrease	Kandasamy and Williams (1984)
Rat	IVT	no effect	Janský (1990)
Angiotensin II			
Rat	IVT	decrease	Janský (1990)[b]
Bombesin			
Mouse	IC	decrease	Janský (1990)
Rat	IVT, H	decrease	Janský (1990)
Cholecystokinin			
Mouse	IVT	decrease	Janský (1990)
Rat	IVT, H	decrease	Janský (1990)
Guinea pig	IVT	increase	Kandasamy and Williams (1983a)

Table 2.5 (*cont.*)

Species	Injection site[a]	Temperature response	References
β-*endorphin*			
Mouse	IVT	increase	Janský (1990)
Rat	IVT, H	increase	Janský (1990)
Met-enkephalin			
Rat	IVT, H	increase	Janský (1990)
α-*melanotropin* (*MSH*)			
Guinea pig	IVT	decrease	Kandasamy and Williams (1984)
Neurotensin			
Mouse	IC	decrease	Prange et al. (1979)
Golden hamster	IC	decrease	Prange et al. (1979)
Rat	IC	decrease	Prange et al. (1979)
Guinea pig	IC	decrease	Prange et al. (1979)
Vasoactive intesntinal peptide			
Rat	IVT	decrease	Janský (1990)
Thyrotropin-releasing hormone			
Rat	IVT, H	increase	Janský (1990)
Arginine vasopressin			
Rat	H	increase	Janský (1990)
Rat	IVT	decrease	Janský (1990)
Cyclic AMP			
Guinea pig	IVT	increase	Kandasamy and Williams (1983a)
Cyclic GMP			
Guinea pig	IVT	decrease	Kandasamy and Williams (1983a)

Note: [a]IVT, intraventricular; AH, anterior hypothalamus; H, hypothalamus; IC, intracisternal.
[b]Janský (1990) presents a review of several studies, and specific sources are cited therein.

"contamination" of microinjection sites with PGE_2 could have dramatic impacts on the thermoregulatory effects of many neuromodulating agents. This may call for a reevaluation of past studies that used microinjection into CNS tissues.

2.4. CNS lesions and temperature regulation

Lesions in the CNS have provided an invaluable tool for understanding its function in humans and experimental animals. Lesions have been especially helpful in studying the cerbral cortex, where there is definitive topographical organization of the sensory and motor outputs. On the other hand, lesion

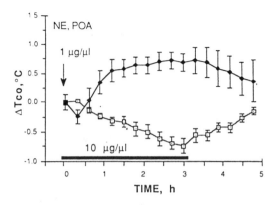

Figure 2.8. Comparison of thermoregulatory responses in unanesthetized guinea pig to administration of NE by microinjection (filled symbols) and by microdialysis (open symbols) for the duration indicated by black bar. Mean core temperature (T_{co}) at time zero was 38.3°C. Reprinted from Quan and Blatteis (1989) with permission from The American Physiological Society.

techniques used for the study of autonomic systems, such as that for temperature regulation, often do not yield definitive results. Searching for a thermoregulatory deficit following a CNS lesion is difficult because autonomic responses can be controlled from multiple levels of the CNS. In many instances, lesioning or sectioning of a supposed central site for thermoregulatory control does not result in a deficit. Moreover, lesions in the hypothalamus, a critical integrative site that exerts control over a variety of autonomic functions, may result in dysfunctions not directly related to thermoregulatory control. This can complicate the interpretation of the true regulatory role of the lesioned site. In spite of these limitations, lesion studies in laboratory rodents (mostly the rat) have given us a breadth of information for characterizing the central neural control of thermoregulation.

Lesion experiments in rodents can be broadly classified into studies on the transient effects and the long-term effects on thermoregulation (Table 2.6). The transient effects of lesioning may be attributable more to the traumatic effects of damage to the CNS than to a given thermoregulatory deficit per se. For example, Rudy and colleagues have characterized a transient febrile response in the rat referred to as *neurogenic hyperthermia* – a regulated rise in core temperature following mechanical damage to the rostral hypothalamic area (Rudy, Williams, and Yaksh, 1977; Ackerman and Rudy, 1980). When the CNS damage occurs at relatively cool ambient temperatures (≤26°C), the hyperthermia is associated with shivering and a reduction in heat loss from the tail. Moreover, the magnitudes of the neurogenic hyperthermia are similar

Table 2.6. *Survey of effects of CNS lesions on changes in thermoregulatory responses of rats*

Lesion site	Effect	Reference
POAH	Hypermetabolism and hyperthermia when maintained at room temperature	Satinoff et al. (1976)
POAH	Impaired autonomic responses for thermoregulation in cold; normal behavioral responses in cold	Satinoff and Rutstein (1970)
POAH	Impaired autonomic responses for thermoreguation at high T_a; near-normal behavioral thermoregulatory responses in heat	Lipton (1968)
POAH	Inability to thermoregulate in hot and cold environments	Van Zoeren and Stricker (1976); Lipton et al. (1974)
POAH, LH/AH	Deficits in salivation and hyperthermia when exposed to high T_a	Toth (1973); Stricker and Hainsworth (1970)
POAH	Reduced/abolished body extension during heat stress	Roberts and Martin (1977)
POAH	Exaggerated amplitude of circadian thermoregulatory rhythm (CTR) lasting for several months	Satinoff et al. (1982)
POAH[a]	Essentially normal autonomic responses to cold stress and febrile challenge	Blatteis and Banet (1986)
POAH[a]	Deficits in peripheral vasomotor tone at cold (but not warm) T_a values	Gilbert and Blatteis (1977)
POAH[b]	Transient elevation in core temperature at cold and warm T_a values	Ackerman and Rudy (1980)
Lateral POAH	Transient hypothermia, vasodilation, and hypometabolism in awake rat	Szymusiak and Satinoff (1982)
Lateral POAH	Altered function of brown adipose tissue (BAT)	Park et al. (1986)
Medial POAH	Transient hyperthermia, vasoconstriction, and thermogenesis in awake rat	Park et al. (1986)
Lateral hypothalamus	Activation of BAT thermogenesis and elevation in body temperature	Corbett et al. (1988)
Paraventricular nucleus	Impaired diet-induced thermogenesis	De Luca et al. (1989)
Substantia nigra	Lowered body temperature, but impaired resistance to heat stress	Brown et al. (1982)
Raphe nuclei	Altered POAH neuronal responses to peripheral thermal stimulation	Werner and Bienek (1990)
Nucleus tractus solitarius	Attenuation in thermogenic responses of BAT to NE and isoproterenol	Fyda et al. (1991b)
Medulla oblongata	Reduced resistance to heat and cold stress	Lipton et al. (1974)
Paramedian reticular nucleus	Reduced resistance to heat, but not cold stress	Lin et al. (1990)

Note: [a]POAH isolated from rest of CNS with microknife cuts.
     [b]POAH damaged by mechanical disruption.

over a wide range of ambient temperatures (10–32°C), suggesting that the hyperthermia is a regulated response, similar to that of a fever. Neurogenic hyperthermia is blocked by administration of antipyretics, which provides strong evidence that the hyperthermia stems from release of PGE_1 and PGE_2 from the traumatized area in the rostral hypothalamus (Rudy et al., 1977).

The transient thermoregulatory effects of a lesion can vary depending on the position of the lesion in the CNS. Using electrolytic lesions of the medial POAH, Szymusiak and Satinoff (1982) found a regulated rise in core temperature similar to that in the mechanical-lesion studies; however, lesions in the ventrolateral POAH led to a regulated fall in core temperature lasting several hours. The transient hypothermic response may well be critical for surviving the trauma to the CNS. Rats given electrolytic lesions in the POAH and kept at a standard room temperature of 25°C experienced mortality greater than 50%. Nagel and Satinoff (1980) discovered that survival following brain lesioning could be greatly improved by maintaining the animals at a cool ambient temperature of 15°C for 12 hr after lesioning. Although the mechanism remains unclear, it was noted that lesioned rats kept at the warmer temperatures had a higher incidence of cardiac arrhythmias. That ambient cooling augments survival following trauma such as CNS lesioning has important implications for therapeutic treatment of trauma and is discussed in more detail in Chapter 9.

After recovery from the initial trauma of a lesion, various thermoregulatory deficits can be detected, depending on the site of injury. Normal thermal homeostasis generally is compromised in the POAH-lesioned rat. Hypothermia and hyperthermia in cold and hot environments, respectively, have commonly been reported (Table 2.6). POAH and lateral hypothalamic lesions result in an inability to salivate when exposed to acute heat stress (Stricker and Hainsworth, 1970; Toth, 1973). Frequently reported observations in rats following electrolytic lesioning of the POAH are severe deficits in autonomic (but not behavioral) thermoregulatory responses (Lipton, 1968; Satinoff and Rutstein, 1970). This provides strong evidence for separate control of behavioral and autonomic processes, with control of the latter under the influence of higher levels of the CNS (e.g., cerebral cortex).

Blatteis and Banet (1986) found that when the POAH of the rat was isolated from the rest of the CNS by means of bilateral knife cuts, autonomic thermoregulatory responses in the cold were essentially normal, as were responses to NE and febrile challenges. In another microknife-cut study it was concluded that a functional POAH was needed for peripheral vasoconstriction (but not thermogenesis) during cold exposure; heat-induced vasodilation was also unaffected by the integrity of the POAH (Gilbert and Blatteis,

1977). In many electrolytic lesions the medial forebrain bundle is destroyed along with the POAH, creating thermoregulatory and other motor deficits not related to the destruction of POAH cell bodies. The microknife studies illustrate the resilient nature of thermoregulatory control: specifically, the ability of other areas to take over thermoregulatory functions when the POAH is isolated from the CNS. The data on electrophysiological, pharmacological, and thermode stimulation discussed in the preceding sections nonetheless substantiate the essential role of the POAH in thermoregulatory control.

2.5. Fever

Much of our understanding of the thermoregulatory systems in humans and experimental animals can be attributed to research on fever over the past century (see Chapter 1). Fever, a malady obviously linked with altered functioning of the thermoregulatory system, is the hallmark of disease, recognized even before the time of the ancient Greek philosophers. The rabbit often has been the preferred experimental model for studying fever in small mammals because of its marked thermal sensitivity to infection. Rodents have not been popular models for fever studies because of early reports that they showed little or no fever during infection. However, careful reassessments of research with rodents over the past few decades have demonstrated unequivocal fevers in response to infection; for reviews, see Kluger (1991), Stitt (1986), and Stitt et al. (1985).

The study of fever is a dynamic field abounding with newly discovered mechanisms; for reviews, see Kluger (1991) and Stitt (1986). The general mechanism of fever in mammals and other species involves aspects of two distinct regulatory processes: the immune system and thermoregulation (Figure 2.9). Basically, exposure to infectious agents or their activating constituents [endotoxins, antigen-antibody complexes, lipopolysaccharide (LPS), components of the cell wall, etc.], termed exogenous pyrogens, stimulates circulating leukocytes and fixed macrophages (e.g., Kupffer cells) to produce endogenous pyrogen (EP), whose major active component is the cytokine polypeptide interleukin 1 (IL-1). Note that an exogenous pyrogen cannot directly cause a fever. Release of EP is an essential step in the development of fever.

Circulating EP appears to reach the POAH via the organovasculosum lamina terminalis (OVLT). The OVLT is a circumventricular organ located in close proximity to the POAH. It allows macromolecules such as IL-1 to pass into the CNS, which otherwise they would not enter because of the blood-brain barrier. The presence of EP in the POAH stimulates conversion of

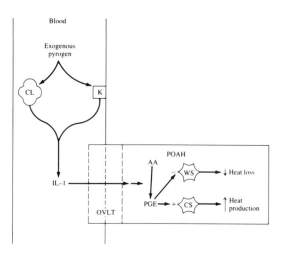

Figure 2.9. Major aspects of the immune and thermoregulatory systems involved in the development of fever in rodents and other mammalian species. CL, circulating lymphocytes; K, Kupffer cell; IL-1, interleukin 1; OVLT, organovasculosum lamina terminalis; AA, arachidonic acid; PGE, prostaglandin E_{1+2}; WS, warm-sensitive neuron; CS, cold-sensitive neuron. Adapted from Boulant (1991), Stitt (1986), and Kluger (1991).

arachidonic acid to PGE_1, PGE_2, and other ecanosinoids. PGE_2 is considered to be a key intermediate for stimulating thermoregulatory responses. The activity and sensitivity of thermosensitive neurons are altered by PGE_2, which leads to an elevation in the set-point. Finally, the elevated set-point causes activation of heat-producing/conserving effectors and inhibition of heat-dissipating effectors, resulting in positive heat storage and an elevation in body temperature.

Antipyresis, or the recovery from fever, has practically evolved into a separate field of study over the past two decades. Drugs used to reduce fever, such as acetylsalicylate (i.e., aspirin) and indomethacin, inhibit the conversion of arachidonic acid to PGE_2. Thus, administration of an antipyretic to a febrile subject causes the set-point to return to normal. At that time the body temperature is above the set-point temperature, and heat-dissipating responses are activated to lower the body temperature to normal. The intervention to treat fever by administering antipyretics is indeed a major research area. Many studies have suggested that the elevated temperature during fever is beneficial to the host and speeds recovery from infection; for a review, see Kluger (1991). The consensus is that the natural recovery from fever involves the release of arginine vasopressin and α-melanocyte-stimulationg hormone

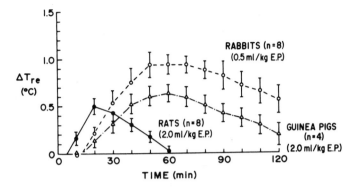

Figure 2.10. Comparison of the effects of intravenous injections of endogenous pyrogen (EP) in rat, guinea pig, and rabbit. Note smaller febrile responses in rat and guinea pig, in spite of larger dose of EP. Reprinted from Stitt et al. (1985) with permission from The American Physiological Society.

(αMSH) into the CNS (Clark and Lipton, 1983; Veale, Cooper, and Ruwe, 1984; Kasting, 1989; Lipton, 1989; Clark, 1991; Kluger, 1991). The release of these polypeptides is most likely involved in preventing the body temperature from reaching dangerous levels during fever.

An outstanding issue in rodent thermal biology is to determine the relative febrile responses to EP and other fever-activating substances. Once it was thought that the relatively large surface area : body-mass ratio of rodents would prevent them from storing enough heat to develop a well-characterized fever. It is not clear why early studies with rats failed to show the presence of fever (Kluger, 1991). Many recent studies have shown that with appropriately controlled conditions, fever can be elicited in rodents with a variety of activating factors: endotoxin, IL-1, PGE_2, and others. For example, a single intraperitoneal injection of *Escherichia coli* endotoxin will elicit a fever in the rat, but the elevation in temperature is apparent only during daylight hours, when body temperature is normally low (Severinsen and Øritsland, 1991). Stitt et al. (1985) compared the febrile responses of the rat, guinea pig, and rabbit to a single, common source of EP (Figure 2.10). They clearly showed a relatively weak febrile response of the rat compared with those in the guinea pig and rabbit.

The low level of fever in the rat may also be related to the structure and function of the OVLT. Compared with that of the rabbit, the OVLT of the rat is not as well vascularized, which may impede the penetration of EP into the CNS during fever (Morimoto et al., 1990). Lesioning the third ventricle and

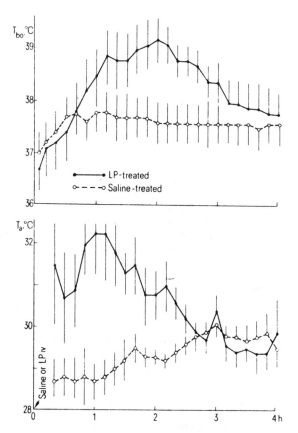

Figure 2.11. Time courses of core temperature (T_{bo}) and selected ambient temperature (T_a) in free-moving guinea pig in a temperature gradient after intravenous administration of saline or 1.0 ml of leukocytic pyrogen (LP) at time zero. Note that the
preference for warmer temperature precedes the elevation in body temperature. Modified from Blatteis and Smith (1980) with permission from Birkhauser Verlag.

OVLT area of the guinea pig results in failure to develop a fever following
LPS injection (Blatteis et al., 1987). Contrarily, lesioning the OVLT region in
the rat and rabbit was found to enhance the febrile response to EP (Stitt,
1985). Although one should be cautious in comparing lesion studies from
different laboratories, it is possible that the febrile mechanisms in rat and
guinea pig may differ considerably.

Blatteis and colleagues have examined several aspects of fever in the
guinea pig. There is a direct relationship between the febrile response to endotoxin and body weight in the rat (Ford and Klugman, 1980) and guinea pig

(Blatteis, 1974), suggesting that surface-area : body-mass relationships could have a role in explaining the relatively weak febrile responses in rodents (Blatteis, 1974). In one of the few studies in which behavioral thermoregulatory responses of rodents were monitored during a fever, guinea pigs given leukocytic pyrogen and placed in a temperature gradient demonstrated a significant increase in the selected ambient temperature, which was closely followed by an elevation in body temperature (Figure 2.11). That the febrile guinea pig prefers a warmer environment along with an elevated body temperature clearly supports the presence of an elevated set-point (i.e., a regulated elevation in body temperature; see Chapter 1).

The numbers of studies of fever in the smaller rodents are meager. Once it was thought that the golden hamster was incapable of generating a fever from administration of LPS. However, using biotelemetric recordings of core temperature to minimize the impact of handling stress on core temperature (see Chapter 5), Conn, Borer, and Kluger (1990) were able to demonstrate a sustained increase in temperature in the hamster following LPS administration. The body temperature in the gerbil monitored with telemetry was also found to increase following LPS, but in this species the LPS caused an unusual reduction in the selected ambient temperature (Akins and Thiessen, 1990). Although they have not been well studied, it appears that mice display very transitory fevers compared with larger rodent species (Kluger, 1991). For example, mice given LPS undertgo a brief 1–2°C elevation in temperature, with a duration of no more than 40 min, followed by prolonged hypothermia lasting for over 2 hr (Habicht, 1981). Akins, Thieesen, and Cocke (1991) recently found that mice in a temperature gradient underwent elevations in selected ambient temperature and body temperature within 90 min after LPS administration.

Interestingly, when the gastrointensidinal bacterial flora are eliminated by administration of nonabsorbable antibiotics in the drinking water, there are slight but significant decreases in the day–night core temperatures of mice and rats (Kluger et al., 1990). Presumably, the normal presence of bacteria in the gastrointestinal tract releases activating substances (i.e., EP) into the circulation, causing a continuous elevation in the set-point. Such a chronic stimulation of febrile processes should surely alter the way we view the control of body temperature in rodents and other species.

A point worthy of consideration in a comparative analysis of fever in rodents is that most studies have been performed at relatively cool room temperatures (e.g., 22°C). The preferred ambient temperatures for small rodents such as the mouse and hamster are 28–32°C (see Chapter 4), values that are likely to increase during fever. Thus, the majority of fever studies in rodents have been performed at ambient temperatures at least 10°C below the projected preferred temperature during fever. The absence of the behavioral thermoregulatory option in these small rodents could seriously alter their normal febrile responses. Future studies should utilize telemetry systems to measure body temperature in unstressed animals and temperature gradients to provide optimal environments for behavioral thermoregulation. Such methods should lead to a better understanding of the comparative aspects of fever in rodents.

3

Metabolism

Metabolism is a universal term in the biological sciences that generally describes the chemical and physical changes in living organisms. In thermal physiology, metabolism almost always refers to the transformation of chemical energy to work and heat (IUPS, 1987). Metabolism can describe various functions, depending on the level of organization. For example, at the cellular and subcellular levels, the heat generated from metabolism is essentially a useless by-product destined for elimination; however, at the whole-animal level that heat is an integral component of thermal homeostasis. In cold environments, metabolic heat is an indispensable thermoregulatory effector. In warm environments metabolism loses its effector status and indeed becomes a liability, adding unwanted heat to the already stressed thermoregulatory system and causing a spiraling elevation in body temperature. Thus, depending on the environmental conditions, metabolism may or may not be considered as a thermoregulatory effector.

The term *metabolic thermogenesis* is commonly applied in situations where the animal's metabolism is utilized as a thermoregulatory effector, such as during exposure to relatively cool environments. To this end, the roles of the principal internal and external factors in the control of metabolism, including body mass, surface area, environmental temperature, insulation, thermal conductance, and motor activity, are discussed in this chapter. The role of metabolism as a thermoregulatory effector during cold exposure is discussed in detail in the next chapter.

3.1. Partitioning of metabolism

The metabolism of the rodent is extremely dynamic, exhibiting marked changes in response to a variety of physical and biological factors, such as ambient temperature and body temperature, wind velocity, altitude,

47

nutritional state, reproductive status, endocrine function, motor activity, and many others. Under any given set of environmental conditions, the total heat produced from a rodent's metabolism is derived from one or more of the following basic sources:

> basal thermogenesis
> postprandial-derived thermogenesis
> shivering thermogenesis
> nonshivering thermogenesis
> diet-induced thermogenesis
> positive-work-derived thermogenesis

Basal or obligatory thermogenesis represents the heat produced from the sum of all anabolic and catabolic biochemical processes involved in the maintenance of respiration, circulation, muscle tonus, peristalsis, body temperature, and other vegetative functions. Postprandial thermogenesis (i.e., specific dynamic action) represents the excess heat production relative to the basal level in the postabsorptive state following the ingestion of food. Shivering thermogenesis and nonshivering thermogenesis are thermoregulatory effectors, representing the principal sources of heat production during short- and long-term cold exposure and fever, and they are covered in the sections on thermoregulatory effectors (Chapter 4) and cold acclimation (Chapter 7). Diet-induced thermogenesis (DIT), a relatively recent discovery in rodent physiology, can represent a significant source of heat production during over-eating and/or ingestion of calorically rich diets. DIT is distinct from postprandial thermogenesis and originates principally from brown adipose tissue (Chapter 4). Positive work (e.g., exercise, increased motor activity) usually is not considered to be a true thermoregulatory effector; however, the heat generated from positive work can represent a significant heat source during thermoregulation in warm and cold environments (see Chapters 4 and 5).

3.2. Methods for measuring metabolism

Metabolism has been measured using both direct and indirect calorimetric methods, the latter being used most frequently in rodent studies (Figure 3.1). Specific details on the theory and operation of direct calorimeters (Benzinger and Kitzinger, 1963; Caldwell, Hammel, and Dolan, 1966; Nagasaka et al., 1979) and indirect calorimeters have been published (Kleiber, 1961; Stock, 1975). Basically, with direct calorimetric devices, the animal is placed in a chamber, and the total nonevaporative or dry heat loss is measured by thermoelectric detectors positioned throughout the inner wall of the chamber. Evaporative heat loss is measured separately by recording the difference in vapor pressures between the influent air and effluent air. The difference in the

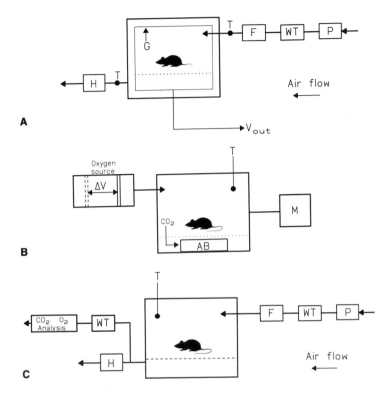

Figure 3.1. Basic principles of operation of calorimeters for measuring metabolism in rodents: direct calorimetry (A), closed-circuit indirect calorimetry (B), and open-circuit indirect calorimetry (C). Key: AB, CO_2 absorbent; F, airflow meter; G, gradient layers of heat sensors; H, humidity or dew-point temperature sensor for measuring evaporative water loss; M, manometer or related device to measure change in pressure in calorimeter; P, pressurized source of air; T, temperature probes for measuring air temperature; WT, water trap to dry air.

temperatures of the influent and effluent airstreams is also determined, which represents a minute fraction of the total dry heat loss. Under steady-state condition, the sum of dry heat loss and evaporative heat loss is equal to the animal's metabolic heat production. With indirect calorimetric methods, metabolism is estimated by measuring the rates of oxygen consumption and/ or carbon dioxide production over a standard time period using open- or closed-circuit respirometers. Open-circuit systems employ a continuous flow of fresh air (usually from a cylinder of compressed air) through a chamber containing the animal. The partial pressure of oxygen (Po_2) is measured in the influent air and effluent air. The change in Po_2 after passing through the chamber multiplied by the airflow rate (STP, dry) yields a measure of oxygen consumption. In closed-circuit systems the chamber containing the animal is

Table 3.1. *Relationships among NPRQ, relative quantities of lipids and carbohydrates metabolized, and heat generated from consuming 1.0 ml oxygen*

NPRQ	Heat (J)	Carbohydrates (mg)	Lipids (mg)
0.707	19.6	0	0.502
0.75	19.8	0.173	0.433
0.80	20.1	0.375	0.35
0.85	20.3	0.58	0.267
0.90	20.6	0.793	0.180
0.95	20.9	1.010	0.091
1.00	21.1	1.232	0

Source: Data from Harper (1975).

hermetically sealed from the outside. A CO_2 absorbent such as a solution of NaOH or ascarite is placed in the bottom of the chamber out of reach of the animal. As CO_2 is produced and absorbed, the air pressure in the chamber begins to decrease, whereupon pure O_2 or air is automatically or manually injected into the chamber at regular intervals to keep the net change in pressure in the respirometer at zero; the injected volume over a given time interval is equal to the oxygen consumed. When ambient air instead of pure O_2 is injected, the chamber must be flushed periodically with ambient air to prevent an excessive decline in the percentage of O_2 in the calorimeter. Details on the technical difficulties of respiratory gas analysis in calorimeters have been published (Bakken, 1991).

The rate of oxygen consumption can be converted to units of heat production by measuring the nonprotein respiratory quotient (NPRQ), which is equal to the ratio CO_2 produced : O_2 consumed, corrected for the gas exchange used in protein metabolism (Harper, 1975):

$$NPRQ = \frac{CO_2 \text{ produced (l)} - ab}{O_2 \text{ consumed (l)} - ac} \tag{3.1}$$

where a is grams of urinary nitrogen produced over the measurement period, $b = 4.75$, and $c = 5.92$. The products ab and ac yield the liters of CO_2 and O_2 produced and consumed per gram of urinary nitrogen derived from protein metabolism, respectively. The NPRQ is rarely measured in rodent metabolic studies; usually it is estimated from values in the literature. Many studies have estimated heat production by assuming an NPRQ of 0.81, and hence 1.0 ml of consumed oxygen is equal to 4.81 calories (cal) or 20.1 joules (J), the latter unit being preferred in conventional metabolic studies. The NPRQ and equivalent heat per milliliter of consumed O_2 are proportional (Table 3.1).

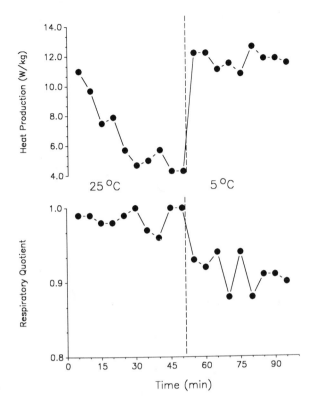

Figure 3.2. Example of how a reduction in ambient temperature causes an increase in metabolic rate and a decrease in respiratory quotient (RQ) in a rat. A decrease in RQ reflects increased use of lipid as a fuel for metabolic thermogenesis during cold exposure. Note the initial rise in metabolism when the animal is first placed in the environmental chamber. Reproduced from Refinetti (1990b) with permission from Springer-Verlag.

Like metabolism, NPRQ is dynamic and can change quickly with varying conditions, such as cold exposure, food deprivation, and other stimuli that alter the proportions of carbohydrates and lipids metabolized (Figure 3.2). Under ad libitum feeding, rats exhibit respiratory quotients (RQs) (i.e., not corrected for protein metabolism) of 0.85 to 1.0, whereas food deprivation lowers the RQ to approximately 0.75 (Table 3.2). Finally, a semantic point should be made regarding various terms mentioned earlier. One often sees the terms *metabolism, metabolic rate, heat production,* and *heat loss* used synonymously. The first three terms are essentially synonymous and are used interchangeably in the literature. In steady-state conditions (i.e., net heat storage is zero), a measure of heat loss provides an accurate measure of heat

Table 3.2. *RQs for laboratory rodents under various intervals of food deprivation*

Species	Time since last feeding (hr)	RQ	T_a (°C)	References
Mouse (NS)[a]	12	0.74	14–35	Herrington (1940)
Hamster	ad lib.	0.8	26	Pasquis et al. (1970)
Rat (NS)	18	0.74	14–34	Herrington (1940)
Rat (SD)[b]	ad lib.	0.87	20	Lackey et al. (1970)
Rat (SD)	24	0.73	20	Lackey et al. (1970)
Rat (SD)	ad lib.	0.97	25	Refinetti (1990b)
Rat (SD)	ad lib.	0.93	5	Refinetti (1990b)
Rat (SD)	20	0.83	25	Refinetti (1990b)
Guinea pig	24	0.77	13–35	Herrington (1940)
Guinea pig	?	0.9	22	Werner (1988)
Guinea pig	ad lib.	0.84	30	Pasquis et al. (1970)

[a]Strain not specified.
[b]Sprague-Dawley.

production. Thus, the terms *heat production* and *heat loss* are often used synonymously in steady-state conditions.

3.3. Basal metabolism

The term *basal metabolic rate* (BMR), a commonly used parameter in clinical metabolic studies, usually is defined as the metabolic rate of an individual that is resting in a thermoneutral state (but is not sleeping) 14–18 hr after eating (Harper, 1975; IUPS, 1987). Theoretically, a major portion of the rodent's metabolism is basal; however, the term BMR is by definition not a practical means of expressing resting metabolism in rodents. One sees the use of BMR as a measure of metabolism in some studies. However, it is quite difficult to create a state in which a rodent is at absolute rest but not sleeping. Indeed, literal application of the term BMR in rodent studies is, in most cases, incorrect. For example, in a thorough analysis of the partitioning of metabolism in the rat, Bramante (1968) found that the rat exhibited periods of no activity for only 4.9% of the 5-hr measuring period. During that inactivity period, the least observed metabolic rate (LOMR) was determined and can be accepted as a measure of BMR for the rat (Table 3.3). However, during nearly half of the time the rat exhibits "microactivities" that correspond to a minimum observed metabolic rate (MOMR). The MOMR is approximately 5% greater than the LOMR, and it was proposed that the former parameter be used as a definition of the rat's resting metabolic rate.

Spontaneous locomotor activity (including feeding and nonfeeding behaviors) accounts for 18–25% of the rat's metabolic expenditure (Morrison, 1968; Brown, Livesey, and Dauncy, 1991). Interestingly, when a rat is maintained at a typical housing environmental temperature of 21°C on a wire-screen floor, its metabolic rate is approximately 25% higher than that at thermoneutrality (ca. 28°C). This additional heat production is derived from both motor activity and a supplemental increase in nonshivering thermogenesis, which wax and wane in reciprocal fashion. That is, when the rat is active, nonshivering thermogenesis is shut down, but during sleep nonshivering thermogenesis increases to replace the heat normally derived from motor activity (Brown et al., 1991). Clearly, an ambient temperature of 21°C in a cage with a wire-screen floor represents a mild cold stress to an individually housed rat.

LOMR and MOMR are used sparingly in rodent studies. Alternatively, the resting metabolic rate (RMR) and standard metabolic rate (SMR) are more applicable and have been measured in rodents under a variety of environmental conditions (Table 3.3) RMR is defined as the metabolic rate of a resting animal that is not in a postabsorptive or fasting state. SMR is the metabolic rate under specified standard conditions, which implies an awake, rested animal that is fasted and under thermoneutral conditions. If any of the aforementioned conditions cannot be met, such as data collected at ambient temperature other than thermoneutral, then the term *metabolic rate* (M or MR) is used as an expression of metabolism.

3.3.1. Allometric considerations: selecting appropriate dimensions

Metabolic rate can be expressed in three general dimensions: energy transformation per unit time (e.g., watts, milliliters O_2 per hour), energy transformation per unit time and body mass (e.g., watts per kilogram, milliliters O_2 per hour per kilogram), and energy transformation per unit time and body surface area (e.g., watts per square meter, milliliters O_2 per hour per square meter). Selecting one or another method for expressing a metabolic rate can be crucial in the interpretation of metabolic data in a rodent study.

It has been recognized since the early part of this century that a species' body mass is a crucial determinant of its metabolism. The classic work of Kleiber and others showed that metabolism in a mammal is controlled primarily by its surface area and body mass. For a brief background, it is well known that the relationship between a homeotherm's body mass (W, g) and metabolism (M, ml O_2 hr^{-1}) is described by a power equation that can

Table 3.3. *Metabolic rate (M) at thermoneutrality and standard room temperature for adult laboratory rodents*

Species[a]	Mass (kg)	M (ml O_2 kg^{-1} min^{-1})	M^b (W kg^{-1})	T_a (°C)	References
Mouse (NS)	0.027	26.1 (SMR)	8.6	33	Herrington (1940)
Mouse (NS)	0.028	44.9	14.8	22	Herrington (1940)
Mouse (A)	0.021	26.8 (SMR)	9.0	26–29[c]	Morrison (1948)
Mouse (OF1)	0.041	29.0	9.7	25	Oufara et al. (1987)
Gerbil	0.065	16.4	5.5	30–35	McRae and Hanegan (1981)
Gerbil	0.065	22.6	7.6	20	McRae and Hanegan (1981)
Gerbil	0.05–0.07	23.7 (RMR)	7.9	32	Luebbert et al. (1979)
Gerbil	0.05–0.07	52.3	17.5	20	Luebbert et al. (1979)
Hamster	0.12–0.14	25.6	8.6	22	Jones et al. (1976)
Hamster	0.098	25.4	8.5	30	Pohl (1965)
Hamster	0.098	29.2	9.8	24	Pohl (1965)
Hamster	0.116	18.1	6.1	26	Pasquis et al. (1970)
Hamster	0.116	19.6	6.6	30	Simek (1976)
Hamster	0.105	39.6	13.3	20	Simek (1976)
Hamster	0.146	14.2 (RMR)	4.9	27–33	Tomasi and Horwitz (1987)
Rat (S)	0.3	17.0 (BMR)	5.7	28	Bramante (1968)
Rat (S)	0.3	17.8 (MOMR)	6.0	28	Bramante (1968)
Rat (SD)	0.39	18.3 (MOMR)	6.1	30	Depocas et al. (1957)
Rat (NS)	0.32	13.4 (SMR)	4.5	31	Swift and Forbes (1939)
Rat (NS)	0.32	22.1 (SMR)	7.4	21.5	Swift and Forbes (1939)
Rat (A)	0.37	14.0 (SMR)	4.6	30	Herrington (1940)
Rat (A)	0.38	17.1	5.6	22	Herrington (1940)
Rat (W)	0.29	21.2	7.1	25	Nagasaka et al. (1979)

Guinea pig	0.53	10.0 (SMR)	3.3	31	Herrington (1940)
Guinea pig	0.63	13.3	4.4	22	Hart and Héroux (1963)
Guinea pig	0.45–0.82	10.8	3.7	22	Werner (1988)
Guinea pig	0.868	13.5 (RMR)	4.7	30	Pasquis et al. (1970)

[a]Parentheses following species indicate genetic strain: S, Simonson; SD, Sprague-Dawley; W, Wistar; A, albino, NS, not specified.
[b]Assuming a metabolic equivalence of 20.1 J per milliliter of O_2.
[c]Bedding material provided.

accurately predict metabolism for species ranging in size from the shrew to the elephant; for reviews, see Schmidt-Neilson (1975a, 1984):

$$M = 3.8W^{0.75} \tag{3.2}$$

The use of logarithms puts the equation in a linear format:

$$\log M = 3.8 + 0.75 \log W \tag{3.3}$$

The fact that the mass exponent of the power equation is 0.75 rather than 1.0 means that metabolism is not directly proportional to body mass. Its value and significance have been debated over the years; for reviews, see Schmidt-Neilson (1984), Heusner (1982), and Hayssen and Lacy (1985). Whereas 0.75 applies to the general shrew-to-elephant curve, calculating the mass exponent for narrowly defined groups, such as species or families of animals under specific conditions, can yield values markedly different from the mainstay 0.75 exponent. For example, comparing various studies of the laboratory rat, values for the mass exponent were found to range from 0.35 to 0.97 (Refinetti, 1989). In spite of these inconsistencies, we find that the RMR data for laboratory rodents in Table 3.3 give a slope of 0.78, a value similar to that in the general mammalian relationship of equation (3.2) (Figure 3.3).

Knowledge of a species' metabolic rate, with no reference to either its body mass or surface area, is of little use in most physiological studies. Commonly, metabolic rate is normalized with body mass (or an exponent thereof) or surface area. For most mammals, the relationship between mass-specific metabolism and body mass (M/W, ml O_2 hr^{-1} g^{-1}) is inverse, with a general relationship of

$$M/W = 3.8W^{-0.25} \tag{3.4}$$

Plotting the watts-per-kilogram metabolic data for laboratory rodents listed in Table 3.3 also yields a relationship with remarkable similarity to the general mammalian relationship of equation (3.4) (Figure 3.3).

The profound effects of body size can be pivotal in selecting appropriate dimensions for rodent metabolic data (e.g., Kleiber and Cole, 1950; Chiu and Hsieh, 1960). Metabolic rate per unit body mass provides a measure of the intensity of tissue metabolism. The marked dependence of this parameter on body mass often can complicate interspecies and intraspecies comparisons. However, calculation of the *metabolic level* (Kleiber and Cole, 1950), which is the metabolic rate normalized to body mass raised to the mass-exponent level (generally 0.75), essentially eliminates the effects of body mass on metabolic rate, as can be illustrated in the calculation of this parameter for the rodent data (Figure 3.3).

Because heat exchange between an animal and its environment occurs over its exposed surface area, expressing metabolic rate in terms of surface area is often preferred. Whereas body mass obviously is a simple parameter to measure, surface area must be estimated using a power relationship, with certain assumptions:

$$SA = kW^{0.67} \qquad (3.5)$$

where SA is surface area in square centimeters, W is the species weight in grams, and the constant k, termed the Meeh coefficient, is species-dependent, with a wide range of intraspecies values, ranging from 6.9 to 13.3 for mouse, 7.2 to 13.0 for rat, and 7.1 to 10.8 for guinea pig (Altman and Dittmer, 1962). These are incredibly wide variations, and they imply that a rodent of a given weight could have as much as double the surface area as another of the same weight, depending on which value of k is used in the calculation. Clearly, there has been considerable error in the determination of surface area (for discussion, see Schmidt-Nielsen, 1984). For rodents such as the mouse, rat, and guinea pig, a k value of 9.0 appears to be most accurate for calculating surface area (Herrington, 1940).

One problem with using surface area as a denominator in metabolic calculations is that the exposed area (i.e., the skin where heat is transferred from the animal to the environment) can vary depending on the animal's behavior. For example, in a hot environment rodents assume a sprawled posture to maximize surface area, whereas in the cold the limbs are pulled close to the body to minimize the exposed surface area (see Chapter 4). Thus, the animal's thermoregulatory behavior affects heat loss per unit of exposed surface area, while heat loss per unit body mass remains relatively constant.

The interplay among body size, surface area, and metabolism can be crucial in various disciplines of the life sciences. Inappropriate selection of metabolic dimensions can produce misleading conclusions. For example, if a given treatment resulted in a 20% decrease in body mass for a 300-g rat, with no other side effects, an increase in metabolic rate from 5.18 to 5.46 W kg^{-1} would be measured in the animal of reduced size. An erroneous conclusion that the treatment caused an elevation in metabolic rate might be reached by a researcher unaware of allometric scaling. On the other hand, calculation of the metabolic level would reveal nearly uniform metabolic rates around 3.83 W kg$^{-0.75}$ for both treatment groups.

A related issue of extreme importance in pharmacokinetic modeling and species-to-species extrapolation is that the dose of a drug or chemical agent to be used in treatment can vary markedly in a species when its metabolism is considered as a dependent variable (Table 3.4). In this theoretical example,

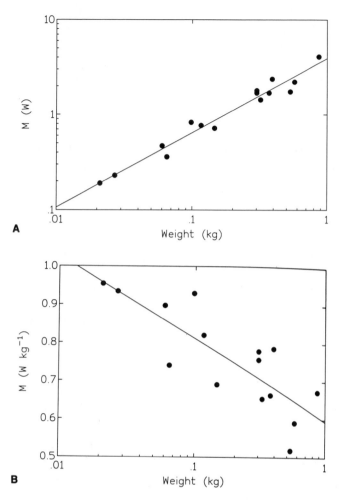

Figure 3.3. Relationships between body weight (*W*) and metabolic rate (*M*) for laboratory rodents maintained at thermoneutrality when the metabolic rate is expressed in different dimensions: (A) heat loss/time (W $= 3.92X^{0.78}$); (B) heat loss/time/body weight (W kg^{-1} $= 3.92X^{-0.21}$); (C) heat loss/time/body weight raised to 0.75 exponent (W kg$^{-0.75}$ $= 3.5 + 0.56X$); (D) heat loss/time/surface area (W m^{-2} $= 42.6X^{0.11}$). Note how calculating the metabolic level (W kg$^{-0.75}$) eliminates the effect of body size on metabolic rate. Metabolic data taken from Table 3.3. Surface areas calculated using equation (3.5) assuming $k = 9.0$.

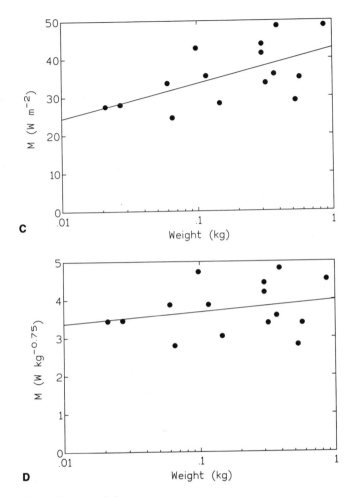

Figure 3.3 (concluded).

I have calculated the doses of a drug for several species using various dimensions of metabolic rate as dependent variables: When a dose of 1.0 mg kg^{-1} is normalized to the metabolic rate it is seen that the 0.02-kg mouse receives only 2 µg W^{-1} kg^{-1}, whereas the 0.8-kg guinea pig receives 28.5 times that dose. Indeed, a 70-kg human would receive 25,000 times that given to the mouse! It is clear that the effects of body mass and surface area on metabolic rate will have marked effects on the turnover of drugs and other chemical agents. This scenario is an obvious oversimplification of drug metabolism and is cited here solely to demonstrate how the expression of

Table 3.4. *Theoretical calculation of doses of a given drug when normalized in various ways to the species' metabolic rates, as calculated from equation (3.3).*

Parameter	Mouse	Hamster	Rat	Guinea pig	Human
Weight (kg)	0.02	0.1	0.3	0.8	70
M (W kg^{-1})	9.9	6.7	5.2	4.1	1.4
M (W)	0.2	0.67	1.56	3.28	98
Drug-treatment doses					
mg kg^{-1}	1.0	1.0	1.0	1.0	1.0
mg	0.02	0.1	0.3	0.8	70
mg W^{-1}	0.1	0.14	0.19	0.24	0.71
μg W^{-1}kg^{-1}	2.0	15	57	195	50,000

metabolism can have far-reaching implications in various disciplines concerned with intraspecies and interspecies extrapolation (see Chapter 9).

3.4. Metabolism during hibernation

Among the rodent species discussed in this book, the golden hamster is the only one capable of true hibernation. At an ambient temperature of 3–5°C a hamster in deep hibernation has a body temperature that is slightly above ambient and a metabolic rate ranging from 0.33 to 0.83 ml (kg^{-1} min^{-1}). This value is approximately 2–5% of the metabolic rate measured at thermoneutrality and meets only 0.7% of the metabolic requirements necessary to remain normothermic in the cold (Lyman, 1948). Clearly, hibernation can afford tremendous energy savings for a cold-exposed homeotherm; for reviews, see Heller (1979) and Lyman (1990).

Along with the dramatic reductions in metabolic rate and body temperature (Lyman, 1948; Kuhnen, 1986), other physiological processes are equally depressed during hibernation: The heart rate is reduced to 4–15 beats per minute; respiration is quite irregular, and the overall rate is approximately 0.5–0.9 breath per minute; the RQ is near 0.7, indicating preferential lipid metabolism; the clotting time of the blood is increased, and the blood pH is near normal (Robinson, 1968). The hamster is considered to be a relatively poor hibernator. It appears that conditions such as age, food availability, and temperature have to be ideal before the hamster will begin to hibernate (Robinson, 1968). Nevertheless, the ability to hibernate provides the hamster markedly enhanced cold tolerance (Anderson, Volkert, and Musacchia, 1971) (see Chapters 2 and 5) and thus continues to serve as a useful experimental model for thermal physiologists.

Table 3.5. *Maximum (peak) oxygen uptake in laboratory rodents when subjected to exercise and/ or exposure to a cold environment*

Species	Mass (kg)	T_a (°C)	Activity status[a]	$\dot{V}_{2\,max}$ (ml O_2 min^{-1} kg^{-1})	References
Mouse	0.026	30	A	132.5	Pasquis et al. (1970)
Mouse	0.034	−10	R	114.2	Pasquis et al. (1970)
Mouse	0.034	−10	A	117.4	Pasquis et al. (1970)
Hamster	0.103	29	A	91.5	Pasquis et al. (1970)
Hamster	0.101	10 to 15	A	117.9	Pasquis et al. (1970)
Hamster	0.098	−25	R	100.0	Pohl (1965)
Rat	0.21	22 to 25	A	110.0	Shellock and Rubin (1984)
Rat	0.21	33 to 35	A	102.0	Shellock and Rubin (1984)
Rat	0.37	24	A	95.1	Shepherd and Gollnick (1976)
Rat	0.22	30	A	76.5	Pasquis et al. (1970)
Rat	0.36	−10	R	50.5	Pasquis et al. (1970)
Rat	0.37	−10	A	59.6	Pasquis et al. (1970)
Guinea pig	0.88	30	A	61.9	Pasquis et al. (1970)
Guinea pig	0.86	−10	R	24.2	Pasquis et al. (1970)
Guinea pig	0.96	−10	A	59.5	Pasquis et al. (1970)

[a]A, active or running; R, resting.

3.5. Maximum (peak) metabolic rate

Maximum metabolic rate (MMR) is the highest metabolic rate sustained during a specified period of aerobic work, whereas peak metabolic rate (PMR) is the highest metabolic rate in a resting state while exposed to a cold environment (IUPS, 1987). To measure the maximum oxygen uptake ($\dot{V}_{O_2\,max}$) in laboratory rodents, combinations of exercise and cold exposure have been used to stimulate heat production (Table 3.5). One must interpret the physiological relevance of $\dot{V}_{O_2\,max}$ carefully because it represents the maximum effort that can be sustained for a relatively short period of time.

There are several key facets to be noted in the $\dot{V}_{O_2\,max}$ data in these rodents (Table 3.5). First, for the animals ranging in mass from the mouse to the guinea pig, $\dot{V}_{O_2\,max}$ is generally equal to five to seven times the metabolic rate at thermoneutrality (Pasquis, Lacaisse, and Dejours, 1970; Hart, 1971). On the other hand, the ratio $\dot{V}_{O_2\,max}$: BMR is about 18 (range, 11–23) for large mammals, including dog, horse, and humans; for review, see Pasquis et al. (1970). Second, the relationship between body mass and $\dot{V}_{O_2\,max}$ in rodents has the same mass exponent (0.75) as does the RMR (Pasquis et al., 1970). Third, $\dot{V}_{O_2\,max}$ in rodents exposed to extremely cold temperatures is generally the same whether or not exercise occurs (Table 3.5). That is, in extreme cold, where the metabolic rate is the PMR, there is a substitution of work-generated heat for heat produced by shivering and nonshivering thermogenesis (Hart and Janský, 1963; Arnold et al., 1986). This does not seem to be true for the guinea pig, which exhibits marked differences in $\dot{V}O_{2\ max}$ while resting and exercising at temperatures of 6°C and −10°C (Table 3.5).

An animal exhibiting its PMR under acute cold exposure is assumed to be mobilizing and utilizing substrates for thermogenesis at its maximum rate. Placing a rodent in conditions to generate its PMR can be a viable way of assessing the limits of catabolic pathways. Wang and colleagues have found that levels of adenosine, a potent antilipolytic agent, may be a major factor in limiting the magnitude of the PMR (Wang and Lee, 1990). The PMR in acute-cold-exposed rats was markedly elevated by preadministration of adenosine deaminase, an enzyme that converts adenosine to inosine. With lowered adenosine levels, the rats were able to mobilize more fatty acids for thermogenesis. Indeed, serum levels of fatty acids in the rat and other mammals increase markedly during acute cold exposure (see Figure 7.2). Paradoxically, adenosine synthesis is increased during cold exposure, a response that is likely to limit the PMR. Developing pharmacological tools to control adenosine metabolism should lead to better ways of improving cold tolerance.

3.6. Metabolic thermoneutral zone

The metabolic rate in rodents and other homeothermic species exhibits three major components as functions of ambient temperature (Figure 3.4). There is a range of temperatures where the metabolic rate is minimal and theoretically is equal to the BMR; this is defined as the species' thermoneutral zone (TNZ). At thermoneutrality, body temperature is regulated primarily by controlling passive heat loss via modulation in the skin blood flow and postural adjustments. These responses require undetectable amounts of metabolic energy, which explains the stable metabolic rate at ambient temperatures within the TNZ. Dry heat loss is essentially minimal at the lower end of the TNZ, meaning that skin blood flow is at basal values. As the ambient temperature is lowered below the TNZ, heat production must be increased to match heat loss in order to maintain a normothermic core temperature. The ambient temperature below which the metabolic rate increases above basal levels is defined as the lower critical ambient temperature (LCT). When the ambient temperature is increased above the TNZ, several behavioral and autonomic processes manifest an elevation in metabolic rate: (1) cellular respiration is

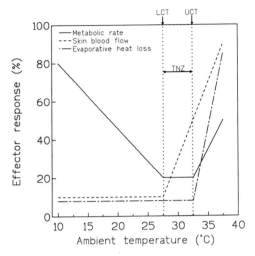

Figure 3.4. Typical relationships between ambient temperature and the activation of principal thermoregulatory effectors: LCT, lower critical ambient temperature, UCT, upper critical ambient temperature, TNZ, thermoneutral zone. Modified from Gordon (1990a) with permission from Pergamon Press.

accelerated directly from a Q_{10} effect[1] of rising body temperature, (2) grooming of saliva on the fur and escape behavior result in increased motor activity and an increased metabolic rate, and (3) the elevation in breathing rate also increases heat production. The ambient temperature at which the metabolic rate increases above the basal levels of the TNZ is often defined as the upper critical ambient temperature (UCT). In some cases the UCT is also defined as the ambient temperature at which evaporative heat-loss mechanisms are activated (IUPS, 1987). Criteria for defining the UCT have not been firmly established. In rodent studies, the metabolic-rate criterion is most widely used, and it is used exclusively in this book.

In rodents, the LCT is a well-studied variable compared with the UCT. In many instances it is difficult to observe a clear elevation in metabolic rate at ambient temperatures above thermoneutrality, whereas the response below thermoneutrality is usually more predictable. It should be remembered that both parameters often are calculated by least-squares interpolation. The metabolic rate at ambient temperatures approximating the LCT or UCT can be quite variable, and a distinct break between basal metabolism and cold- or

[1]Q_{10} is the ratio of the rate of a physiological process (R_2) at a given temperature (T_2) to its rate (R_1) at a temperature (T_1) that is 10°C lower. If the temperatures are not separated by 10°C, Q_{10} can still be calculated by the formula $Q_{10} = (R_2 - R_1)^{10/(T_2 - T_1)}$. See Schmidt-Nielsen (1975b).

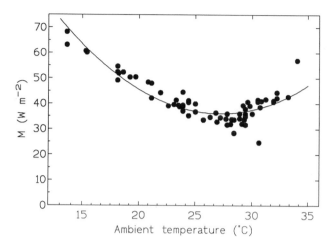

Figure 3.5. Example of the relationship between ambient temperature and metabolic rate in the albino rat. Note the variation in metabolic rate around the TNZ, which was estimated to be 29–30°C. The regression line represents a third-order polynomial calculated by the author. Data from Herrington (1940).

heat-induced increases in metabolism can be difficult to discern (Figure 3.5). In spite of this, the UCT and LCT are useful for comparing overall thermoregulatory sensitivity between species.

A species' TNZ can be affected by a variety of internal and external factors, including sleep/wake pattern, health status (e.g., fever), body mass, gender, wind speed, and others. For the most part, the TNZ data for most rodents have been determined in resting animals while attempting to control or eliminate many of the aforementioned factors. Not surprisingly, there are some interspecies as well as intraspecies variations in the TNZ characteristics of laboratory rodents (Table 3.6). The reported TNZs are in some cases less than 1.0°C (rat) and as high as 14°C (guinea pig). Data for the rat are relatively abundant compared with other rodents species. The LCT values for rat, hamster, gerbil, and mouse have been relatively consistent across most studies with values of 28–30°C. On the other hand, the LCT data for the guinea pig fall into two groups, with values of either 20–21°C or 29–30°C. The reason for this disparity is not clear. Procedural differences involving training, adjustment time in the chamber, level of activity, age, and other variables may account for the variation in this species.

3.6.1. Metabolic rate below thermoneutrality

Other than the disparity of the guinea pig data mentioned earlier, it is clear that the LCT in laboratory rodents is significantly higher than the typical

Table 3.6. *Survey of studies of LCT and UCT in laboratory rodents under resting conditions, with metabolic rate between LCT and UCT presumed to be minimal or basal*

Species	LCT (°C)	UCT (°C)	References
Mouse (NS)[a]	30.6	~34	Herrington (1940)
Mouse (OF1)	26	30	Oufara et al. (1987)
Mouse (BALB/c)	31	34	Gordon (1985)
Gerbil	30	35	McRae and Hanegan (1981)
Gerbil	30	39	Robinson (1959)
Gerbil	28	32	Randall and Thiessen (1980)
Gerbil	32	34	Luebbert et al. (1979)
Hamster	29	29	Pace and Rahlman (1983)
Hamster	<30	—[b]	Pohl (1965)
Hamster	28	34	Gordon et al. (1986)
Rat (NS)	29.2	31.0	Herrington (1940)
Rat (NS)	26.5	26.5	Pace and Rahlman (1983)
Rat (NS)	30	33	Swift and Forbest (1939)
Rat (HZ)	28	>33	Clarkson et al. (1972)
Rat (W)	28	32	Poole and Stephenson (1977a)
Rat (SD, F344)	28	30–32	Gordon (1987)
Rat (LE)	28	32–34	Gordon (1987)
Rat (SD)	22	27	Gwosdow et al. (1985)
Guinea pig	21	29	Hart (1971)
Guinea pig	30	31	Herrington (1940)
Guinea pig	29	29	Pace and Rahlman (1983)
Guinea pig	20	34	Gordon (1986)

[a]Genetic strain: HZ, Holtzman; W, Wistar; SD, Sprague-Dawley; LE, Long-Evans; F344, Fischer 344; NS, not specified.
[b]Not reported.

room temperature in most laboratory settings (e.g., 20–24°C). Thus, when working with laboratory rodents under normal ambient conditions, their metabolic rates likely will be significantly elevated above basal levels (see Table 3.3). This can confound the interpretation of a variety of nutritional, pharmacological, and toxicological studies in which the assumption of a minimum metabolic rate is critical.

The metabolic rate at temperatures below the TNZ can be described using a simple linear relationship derived from Fourier's law of heat flow and Newton's law of cooling (Kleiber, 1972a):

$$M = C'(T_b - T_a) \tag{3.6}$$

where the rate of heat loss (M, assuming that heat loss equals heat production) is directly proportional to whole-body thermal conductance (C') and to the difference between body and ambient temperatures ($T_b - T_a$). The pivotal work of Scholander et al. (1950) demonstrated that this equation models the metabolic responses of a variety of mammals exposed to temperatures below the LCT. It can be seen that if the metabolic rate is extrapolated to zero (see dashed line in Figure 3.4), then T_b will equal T_a. In other words, if it were not for basal metabolic processes, the metabolic requirements for thermoregulation at an ambient temperature of 37°C theoretically would be zero. The principle of equation (3.6) can be useful for comparing thermoregulatory sensitivities between rodent species and for providing approximate values for metabolic rates at any temperature below the LCT. That is, if one knows the thermal conductance (discussed later) and assumes a core body temperature of 37°C, equation (3.6) will yield metabolic rates that in many instances will be close to the observed values at ambient temperatures below the TNZ (Table 3.3).

How well do the overall metabolic responses of laboratory rodents to ambient cooling follow the linear function of equation (3.6)? In reality, the metabolic rate often may exhibit a nonlinear elevation during cooling. Researchers mindful of the Scholander relationship, or for want of a simplified analysis, often will try to fit a linear function to their data in spite of nonlinearity in the animal's metabolic response to cooling. Some studies have reported that the metabolic rate below the LCT does indeed follow the ideal linear-response model (Figure 3.6). On the other hand, there are notable numbers of deviations from the ideal response, especially in studies of rats (Gordon, 1990a). This is illustrated in the extrapolation to a metabolic rate of zero, which in many studies of rats has deviated markedly from 37°C. Of course, one should be wary in comparing studies from different laboratories as is done in Figure 3.6 because there are many uncontrolled or unreported variables that could affect the temperature–metabolic-rate relationship. At this point it is sufficient to conclude that until a thorough interspecies comparison of metabolic responses to ambient cooling is completed, the response of the rat appears to deviate considerably from those of other rodent species.

3.7. Physical factors affecting metabolism

3.7.1. Thermal conductance (whole body)

Thermal conductance, which can be defined as the facility or ease of heat loss to the environment, is an important parameter in rodent thermal biology (Hart, 1971; Bradley and Deavers, 1980; Aschoff, 1981; Lovegrove, Heldmaier, and Ruf, 1991). Thermal conductance essentially determines the metabolic requirements for an endotherm to remain normothermic. That is, the

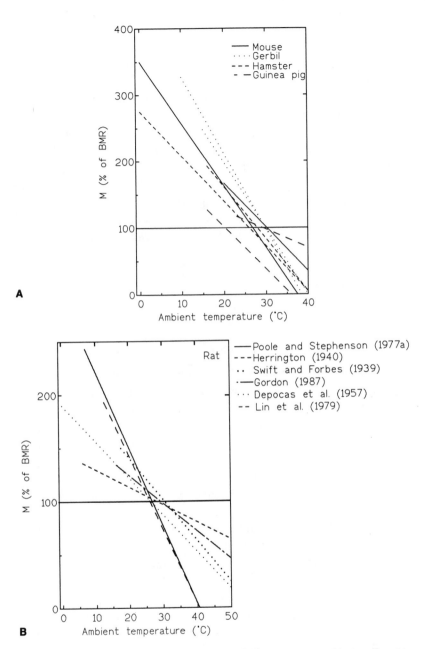

Figure 3.6. Plots of various species' metabolic responses to ambient cooling at temperatures below LCT for mouse, hamster, gerbil, and guinea pig (A) and rat (B). Note that in many studies the regression lines do not intersect with normal body temperature when extrapolated to a metabolic rate of zero. Sources: (A) Robinson (1959), Pohl (1965), McRae and Hanegan (1981), Pace and Rahlman (1983), Gordon (1985, 1986), Gordon et al. (1986); (B) modified from Gordon (1990a).

higher the thermal conductance, the greater the metabolic requirements at a given ambient temperature below the LCT. Thus, thermal conductance is a crucial variable that limits the environmental extremes that an endotherm can tolerate and still remain normothermic (Lovegrove et al., 1991).

Thermal conductance is dependent on several physical, physiological, and behavioral factors: ambient temperature, insulative quality of the fur, skin blood flow, motor activity, and sleep–wake cycle. In physical terms, thermal conductance is defined as "the rate at which heat is conducted between unit area of two parallel surfaces in a medium when unit temperature difference is maintained between them" and has dimensions of watts per square meter per degree Celsius (IUPS, 1987). This definition is infrequently applied in rodent studies, although the term "thermal conductance" is commonly used. In many studies, thermal conductance is measured in terms of body mass rather than surface area, and to avoid confusion with the foregoing physical definition it is often termed "whole-body thermal conductance." Its value (C') can be derived from equation (3.6) (also see Chapter 1):

$$C' = M/(T_b - T_a) \tag{3.7}$$

It will be noted that the value C' is the slope of the ambient-temperature–metabolic-rate regression line below the LCT (see Figure 3.4). Thermal conductance (whole body) is often reported in units of milliliters O_2 per minute per kilogram because of the predominant use of indirect calorimetry in measuring metabolism. However, when possible it is preferable to express thermal conductance in terms of heat transfer rather than O_2 consumption (e.g., watts per kilogram per degree Celsius).

It is generally found that thermal conductance is minimal and constant at ambient temperatures below the LCT (McNab, 1980). Thus, thermal conductance remains constant as metabolism increases with decreasing ambient temperature. Raising the temperature above the LCT leads to an elevation in thermal conductance because of peripheral vasodilation and increased evaporation. Below the LCT, evaporation accounts for 5–25% of total heat loss (see Chapter 4). If evaporation is not taken into account, then the calculation of "wet thermal conductance" is appropriate and has been used widely in comparisons of rodent species maintained at temperatures below the LCT, as discussed later. On the other hand, wet thermal conductance is meaningless at ambient temperatures above the TNZ, where evaporation is a significant avenue of heat loss (McNab, 1980).

Body mass plays a major role in governing the value of thermal conductance (Bradley and Deavers, 1980; Aschoff, 1981). Aschoff (1981) developed an allometric equation predicting thermal conductance for resting mammals

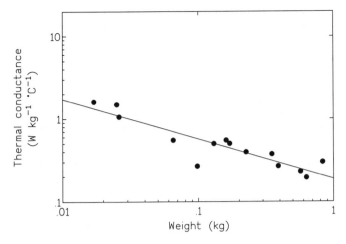

Figure 3.7. Effects of body mass on whole-body thermal conductance in mouse, gerbil, rat, and guinea pig. Thermal conductance measured in animals exposed to ambient temperatures below thermoneutrality. Data from Aschoff (1981), Bradley and
Deavers (1980), Hart (1971), McRae and Hanegan (1981), Gordon (1985, 1986),
Gordon et al. (1986).

ranging in body weight from 0.004 to 6.6 kg maintained at temperatures below the TNZ:

$$C' \ (\text{ml O}_2 \ \text{g}^{-1} \ \text{hr}^{-1} \ °\text{C}^{-1}) = 1.022W^{-0.519} \tag{3.8}$$

Or, assuming an RQ of 0.81, C' can be calculated in calorimetric units of
heat transfer:

$$C' \ (\text{W kg}^{-1} \ °\text{C}^{-1}) = 0.158W^{-0.519} \tag{3.9}$$

Thus, thermal conductance decreases dramatically with increasing body size.
This is attributed in part to the inverse relationship between body mass and
metabolic rate (see Figure 3.3). However, the slope of the curve of body mass
versus thermal conductance is more than double that for the body-
mass–metabolic-rate relationship; see equation (3.6). In other words, although the calculation of C' is derived from metabolic rate, it is clear that the
effects of body size on metabolic rate do not fully explain the slope of the
curve of body mass versus thermal conductance. There clearly are intraspecies and interspecies characteristics other than body size that contribute to
the allometric scaling of thermal conductance.

Thermal conductance in common laboratory rodents follows an inverse dependence on body size, similar to the general mammalian relation of equation
(3.8) (Figure 3.7). The slope of the plot of the regression equation has been

shown to be unique to particular mammalian orders and families (Bradley and Deavers, 1980) and can change dramatically between periods of activity and inactivity (Aschoff, 1981), torpor and hibernation (Snyder and Nestler, 1990).

3.7.2. Tissue thermal conductance

Tissue thermal conductance is defined as "the rate of heat transfer per unit area during steady state when a temperature difference of 1°C is maintained across a layer of tissue" (IUPS, 1987). This term is often used to measure the heat transfer between a tissue and its immediate environment, such as from a tissue or organ to the blood stream or from the body core to the peripheral tissues and skin. In this latter case, tissue thermal conductance is calculated as

$$K = M - E_{res}/T_r - T_{sk} \qquad (3.10)$$

where K is tissue thermal conductance (W m^{-2} °C^{-1}), M is metabolic rate (W m^{-2}), E_{res} is respiratory evaporative water loss (W m^{-2}); T_r is rectal or core temperature (°C), and T_{sk} is mean skin temperature (°C).

This parameter is used infrequently in rodent studies because of the difficulty in measuring T_{sk} and E_{res} (see Chapter 4). Such data would normally be collected from restrained animals, which likely would be under considerable stress. In spite of these limitations, there have been some measurements of tissue thermal conductance in restrained rats yielding minimum values of 6–7 W m^{-2} °C^{-1} at temperatures below the TNZ to as high as 16 W m^{-2} °C^{-1} during heat stress (Lin et al., 1979; Collins, Hunter, and Blatteis, 1987). Values of tissue thermal conductance for rodents can be useful in studies of heat transfer. As with whole-body thermal conductance, the rise in tissue thermal conductance reflects peripheral vasodilation as heat is transferred from the core to the cooler exterior surfaces to enhance heat dissipation. To develop concise models of heat transfer in laboratory rodents it will be necessary to collect more data on tissue thermal conductance, preferably in unrestrained animals.

3.7.3. Pelt insulation

Other than body size and surface area, the insulative quality of the pelt is one of the few physical factors that affect the metabolic rate. Unfortunately, the insulative qualities of the pelt have not been seen as a major issue in rodent thermal biology. One reason for the meager data base may be that small mammals generally show relatively little change in pelt insulation between summer and winter, as compared with larger species (e.g., Folk, 1974) (see Chapter 7). Yet the fur is nonetheless a critical component of thermal homeostasis in rodents.

Fur accounts for 1.6–2.1% of body mass in the adult rat (Joy, Knauft, and Moyer, 1967; Roussel and Bittel, 1979), but only ~0.2% in the adult mouse (Al-Hilli and Wright, 1988a). Depilation results in a 50% elevation in metabolic rate for rats maintained at 24°C (Roussel and Bittel, 1979) and a 7°C increase in the threshold ambient temperature for initiation of shivering (Stoner, 1971). At thermoneutrality, the metabolic rate for a strain of genetically hairless mice is 25% above that for normally furred mice; at 22°C the metabolic rate is 51% greater in the hairless animals (Mount, 1971). The insulative capacity of excised mouse pelt can be increased by approximately 25% by ruffling the fur, that is, to simulate piloerection during cold exposure (Barnett, 1959). Thus, the insulation provided by the pelt is not the most dynamic thermal physiological variable, but it is nonetheless a major determinant of a species' metabolic rate and thermoregulatory capacity. The role of pelt insulation in cold acclimation is discussed further in Chapter 7.

4

Thermoregulatory effector responses

Rodents and other homeotherms utilize a variety of effectors (i.e., motor outputs) to maintain a stable body temperature over a relatively wide range of environmental temperatures. This is achieved through the integration of behavioral and autonomic neural processes under the control of key sites in the CNS, particularly within the rostral brain stem (see Chapter 2). When the subject is maintained at a thermoneutral temperature, thermoregulation in the resting endotherm is achieved through a balance between (1) the heat liberated from basal metabolic processes and (2) modulation of peripheral vasomotor tone (i.e., skin blood flow). Lowering or raising the ambient temperature beyond the thermoneutral zone (TNZ) causes the activation of heat-gain/conserving or heat-dissipating effector responses, respectively (Figure 4.1).

It is reasonable to assume that a major portion of a rodent's life is lived in environments that are not thermoneutral. This is particularly true for most laboratory rodents, whose housing and experimental conditions are thermostabilized for human comfort, but happen to be markedly below the TNZ for most rodents (see Chapter 3). This chapter discusses the properties of the autonomic and behavioral thermoregulatory effectors utilized by laboratory rodents during transient exposures to temperatures above and below thermoneutrality. Their thermoregulatory responses to prolonged exposure to heat and cold stress are treated separately in the discussion of temperature acclimation (see Chapter 7).

4.1. Peripheral vasomotor tone

The summation of conductive, convective, and radiative heat losses, commonly referred to as dry heat loss and less frequently as sensible or Newtonian heat loss (IUPS, 1987), is controlled via adjustments in peripheral

73

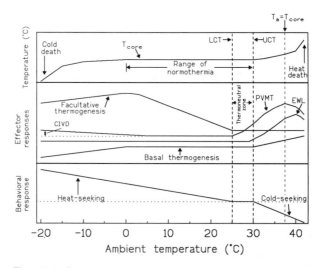

Figure 4.1. General relationships between ambient temperature and the onset and relative magnitudes of thermoregulatory effectors in a typical nonhibernating endotherm: LCT, lower critical ambient temperature; UCT, upper critical ambient temperature; EWL, evaporative water loss; BMR, basal metabolic rate; CIVD, cold-induced vasodilation; PVMT, peripheral vasomotor tone. A dotted line indicates the basal value for each variable. Facultative thermogenesis includes heat produced from shivering and nonshivering thermogenesis.

vasomotor tone (PVMT). Dry heat loss is increased when blood flow from the warm core is directed to the cool peripheral tissues. This is achieved via central neural modulation of smooth-muscle tone in both arteriolar precapillary sphincters and arteriovenous anastomoses (AVAs). Dry heat loss is most efficient from uninsulated, sparsely furred surfaces such as the eyes, nose, ears, feet, and tail. Control of PVMT involves not simply thermal stimuli but also complex integration of other factors, including blood pressure, tissue O_2 availability, and level of activity (Guyton, 1986). Changes in any one of these factors can lead to marked effects on PVMT and heat loss.

4.1.1. PVMT reflexes of tail

The thermoregulatory system in rodents has evolved a variety of morphological and physiological adaptations for utilizing PVMT to control dry heat loss. Among the distinct adaptations to be discussed, the tail of the rat is by far the best-understood structure. Several physiological and anatomical characteristics contribute to make the tail a crucial site for heat exchange: (1) it lacks fur, (2) it is well vascularized, with a high density of AVAs (Gemmell and Hales, 1977), permitting a high rate of blood flow during heat stress

(Figure 4.2), and (3) it has a relatively high surface area : volume ratio, accounting for approximately 7% of the total surface area (Lin et al., 1979), which further enhances heat dissipation. At birth, the tail of the rat is essentially devoid of noradrenergic nerve terminals (Anderson and McLachlan, 1991). Sympathetic innervation of the vascular aspects of the tail develops rapidly over the first 6 weeks of life. This postnatal development of the noradrenergic innervation of the tail suggests that its vasomotor control should be amenable to environmental influences, such as cold or warm acclimation (see Chapters 6 and 7). Details on the vascular (Thorington, 1966) and neural anatomy (Anderson and McLachlan, 1991) of the rodent tail have been published.

The tail in both rat and mouse has been found to be crucial for thermoregulation. For example, amputating the tail reduces heat tolerance and results in higher body temperatures following administration of thermogenic drugs (Harrison, 1958; Stricker and Hainsworth, 1971; Spiers, Barney, and Fregly, 1981). The rate of growth of the tail in immature mice and rats has been found to be reduced during cold acclimation (Chevillard, Porter, and Cadot, 1963; Barnett and Widdowson, 1965). In a cold environment, a shorter tail, with less surface area, aids in reducing dry heat loss, whereas a longer tail in a warm environment facilitates heat loss (see Chapter 7).

PVMT in the tail is exquisitely sensitive to changes in ambient temperature. At standard room temperatures (e.g., 20–25°C), blood flow to the tail of the rat is near zero; as the temperature is increased above threshold there is a marked, exponential rise in blood flow, reaching a value more than 10 times the tail's basal level (Rand, Burton, and Ing, 1965). When the tail is vasodilated, blood flow is directed predominantly from the ventral artery to the lateral veins (Figure 4.2A) and through both capillary vessels and AVAs (Richardson, Shepherd, and McSorley, 1988). Control of blood flow through the AVAs is mediated through changes in sympathetic constrictor tone (Richardson, Qing-Fu, and Shepherd, 1991). When ambient temperature is held constant, a sudden rise in temperature in the tail is indicative of increased blood flow. However, because of the ventral-to-lateral flow of blood, one must be cautious in the placement of temperature probes on the tail. Cooler temperatures are likely to be measured on the dorsal aspects of the tail when it is vasodilated (Figure 4.2A).

The threshold ambient temperature for vasodilation in the tail of a restrained rat ranges between 27°C and 30°C (Table 4.1). Under steady-state conditions the tail can dissipate approximately 25% of a rat's basal metabolic heat production at ambient temperatures approximating the LCT (Young and Dawson, 1982). Of course, the effectiveness of the tail in dissipating heat is

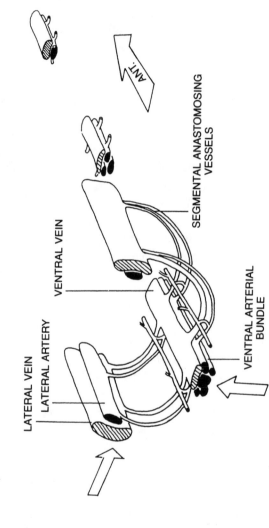

Figure 4.2A. Diagram of the vasculature of the rat tail. Note that the relatively small caliber of the ventral vein as compared with the lateral vein and the large caliber of the ventral artery as compared with the lateral artery suggest a predominance of blood entering the tail ventrally and exiting laterally. Reproduced from Young and Dawson (1982) with permission from the National Research Council of Canada.

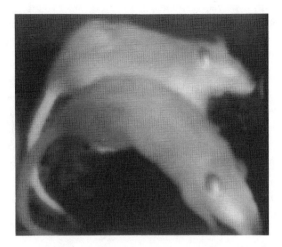

Figure 4.2B. Demonstration of heat loss from the tail of the albino rat using infrared thermography (lighter colors indicate warmer surface temperatures). This is a thermogram showing two rats maintained at an ambient temperature of 26.4°C. The rat at the top of the thermogram was untreated. Note the bright color of its tail, which was a result of tail vasodilation. The bottom rat had been given NE intramuscularly at 1 mg kg^{-1}, 30 min prior to taking the thermogram, to promote peripheral vasoconstriction. Note the dark color of the tail, which is indicative of low skin temperature and reduced blood flow. Infrared thermogram provided courtesy of Professor J. E. Heath and J. Wang, University of Illinois.

Table 4.1. *Threshold ambient temperatures for vasodilation ($T_{a\text{-}vd}$) of the tail and foot in restrained laboratory rodents*

Species	Appendage	$T_{a\text{-}vd}$ (°C)	References
Rat	tail	28	Hellström (1975a,b)
Rat	tail	27–30	Rand et al. (1965)
Rat	tail	30	Dawson and Kleber (1979)
Rat	tail	27	Clarkson et al. (1972)
Rat	tail	22	Lin et al. (1979)
Rat	foot	22	Lin et al. (1979)
Hamster	foot	38	Reigle and Wolfe (1974)

proportional to the skin–ambient-temperature gradient; hence, heat loss is attenuated as ambient temperature is elevated. It is interesting to note that blood flow in a vasodilated tail is quite dynamic, showing rhythmic, pulsatile flow. For example, when a rat is maintained at a suprathreshold temperature of 29–33°C, its tail exhibits oscillatory on–off periods of vasodilation and vasoconstriction, with a period of approximately 20 min (Young and Dawson, 1982). Oscillations in skin temperature, which generally prevail in most endothermic species, may be critical in thermoregulatory stability (Gordon and Heath, 1983).

Under most environmental scenarios, increased blood flow to the tail will facilitate heat loss as tail skin temperature rises above the temperature of its surroundings. However, if the environmental temperature exceeds tail skin temperature, then increased blood flow will *add* to the body heat burden as the tail absorbs heat from the ambient air. Raman, Roberts, and Vanhuyse (1983) found that tail blood flow as a function of tail temperature reached its maximum at 35°C, and then decreased with further increases in temperature (Figure 4.3). This response reduces heat gain from the environment and appears analogous to the PVMT of human extremities subjected to peripheral heating (Raman et al., 1983).

In anesthetized rats maintained at a core temperature of 39.5°C, increasing the tail skin temperature above 36°C also causes a reduction in tail blood flow (Sakurada, Shido, and Nagasaka, 1991). Interestingly, the heat-induced vasoconstriction reaches a nadir at approximately 41°C, but then increases as skin temperature is elevated to 44°C. This response, which occurs in the denervated tail as well as the intact tail, may represent another level of thermoregulatory protection. That is, at 44°C, blood in the core is approximately 4°C lower than that in the tail skin, whose temperature is precipitously close to causing irreversible cell damage. Thus, increasing tail blood flow will in fact "cool" the tail, providing protection from heat damage.

The activation of vasodilation of the tail with increasing ambient temperature is mediated with little or no change in core temperature in the awake rat. Thus, it can be assumed that such responses are mediated predominantly by the activity of peripheral thermal receptors. On the other hand, elevations in core temperature, as will occur with administration of metabolic stimulants, exercise, or artificial heating of the core, can also induce PVMT responses in the tail. The threshold core temperature for vasodilation of the rat tail ranges from 37°C to 39.8°C; for review, see Gordon (1990a). Although it has not been well studied in rodents, it seems likely that this threshold core temperature could be modulated by raising or lowering ambient temperature (see Chapter 2).

Data on tail PVMT in other rodent species are meager. The tail of the mouse has been shown to experience an increase in blood flow in response to internal heating induced with radio-frequency (RF) radiation (Gordon, 1983a). Otherwise, little is known of the physiological control of PVMT in this species. The Mongolian gerbil has a relatively long tail, but shows little, if any, increase in blood flow during ambient heating (Klir, Heath, and Ben-

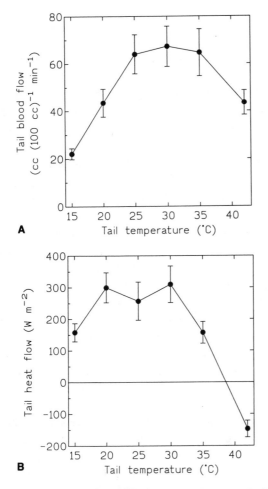

Figure 4.3. Relationships between tail temperature and tail blood flow and heat flow in a restrained rat whose core temperature was maintained at 39°C. Maximum blood flow peaked at a tail temperature of 30–35°C and then decreased with further tail heating. Note negative heat flow (i.e., net increase in heat to the tail) at a tail temperature of 42°C. Data from Raman et al. (1983).

nani, 1990). In studies with infrared (IR) thermography it was found that this rodent has little control over any of its surface temperatures over a relatively wide range of ambient temperatures (−10°C to 35°C). This apparent lack of thermoregulatory-mediated PVMT reflexes could make the gerbil a very interesting model to study the integration of cardiovascular and thermoregulatory mechanisms.

4.1.2. PVMT of feet and ears

The feet and ears are well-vascularized, bare surfaces that can serve as "windows" for heat dissipation in some rodents. The ears of the guinea pig have AVAs and will show an active vasodilation when core temperature exceeds 39°C (Grant, 1963). During recovery from hyperthermia, the pinna of the guinea pig shows an abrupt constriction at a core temperature of 39.2°C (Szelenyi and Hinckel, 1987). Otherwise, little is known of the thermoregulation of PVMT in this species. Interestingly, the rat ear lacks AVAs and shows no vasodilation in response to whole-body heating or exercise (Grant, 1963; Thompson and Stevenson, 1965). The pinna of the gerbil also shows little, if any, thermoregulatory-mediated changes in PVMT (Klir et al., 1990).

The feet and distal portions of the limbs of the rat make up approximately 10% of its total surface area, slightly more than the surface area of the tail (Lin et al., 1979). The foot vasodilates in near synchrony with the tail during internal heating caused by exercise or by treatment with hyperthermia-inducing drugs, as well as during ambient heat stress (Thompson and Stevenson, 1965; Lin et al., 1979; Gordon and Watkinson, 1988). Unfortunately, it is difficult to obtain recordings of skin temperatures in the feet of unstressed, free-moving rats using conventional probes. However, with IR thermographic techniques, such measurements could be made with ease and would add greatly to our understanding of these sites in the control of heat exchange (Klir et al., 1990).

A change in whole-body thermal conductance can serve as an indication of an alteration in PVMT (see Chapter 3). Thermal conductance increases as ambient temperature rises above the LCT. The ambient temperature needed to increase thermal conductance represents a threshold for a change in PVMT and can be another means of comparing the sensitivities of dry heat losses in unrestrained species. For example, threshold ambient temperatures for increases in thermal conductance in the absence of any marked change in evaporative water loss are 25°C for mouse (Gordon, 1985), 28–30°C for golden hamster (Gordon et al., 1986), 30–32°C for the Fischer rat (Gordon, 1987), and ~30°C for guinea pig (Gordon, 1986).

4.1.3. Cold-induced vasodilation

When the tail of a rat is immersed in a bath of crushed ice, the skin temperature of its tail drops precipitously to near 0°C and then shows a wave of vasodilation, reaching a peak temperature of 1.6°C within 3.2 min after immersion (Hellström, 1975b). These waves of cold-induced vasodilation (CIVD), which appear to be related to the classic Lewis "hunting response" in cold-exposed human fingers, will continue for up to 30 min. In the ice-bath immersion model, the incidence of CIVD increases linearly with ambient temperature, suggesting a major role of peripheral thermal afferents in the modulation of CIVD. Interestingly, CIVD can also be induced in isolated sections of the rat's tail artery; however, the period of such CIVD is much longer than that in the intact animal (Gardner and Webb, 1986). These in vitro studies suggest that cold-induced vasoconstriction is associated with enhanced smooth-muscle responsiveness to norepinephrine, whereas CIVD occurs because of a cessation in the secretion of vasoconstrictive neurotransmitters in the tail.

4.2. Metabolic (facultative) thermogenesis

The metabolic requirements for endotherms can be divided into obligative and facultative processes (Himms-Hagen, 1986). Obligative metabolism, or the basal metabolic processes, suffices for thermal homeostasis when the animal is exposed to temperatures in the TNZ (see Chapter 3). Facultative processes include the metabolic responses that must be recruited to maintain thermal homeostasis as ambient temperature is reduced below the TNZ, and they are derived from two principal effectors: shivering thermogenesis and nonshivering thermogenesis.

As discussed in Chapter 3, small mammals with high surface-area : body-mass ratios are faced with the problem of increased heat loss in the cold and must utilize facultative thermogenesis to a much greater extent than larger species. Moreover, because of their small size, rodents are unable to grow a thick insulative pelt, thus placing further demands on their facultative metabolic requirements (for further discussion, see Chapters 3 and 7).

4.2.1. Shivering thermogenesis

Large mammals, including humans, rely primarily on shivering to thermoregulate in the cold (Folk, 1974; Hensel, 1981). Rodents also utilize shivering to generate heat, but when faced with continuing cold stress, nonshivering thermogenesis replaces shivering as their primary source of heat production, as discussed later. This transition is thought to be beneficial from several standpoints: (1) the heat from shivering is generated peripherally, resulting in relatively inefficient warming of the core tissues; (2) shivering is an uncomfortable state, and the skeletal muscles involved in shivering cannot be used simultaneously as well for other motor functions; (3) body movements caused by shivering can increase heat loss, an effect that appears to be augmented in relatively small species, further reducing the efficiency of shivering (Janský, 1979; Kleinebeckel and Klussmann, 1990). An interesting historical note is that for many years it was thought that shivering was the only source of heat for facultative thermogenesis. It was not until 1955 that Davis and Mayer first reported a distinction between shivering and nonshivering thermogenic mechanisms; for review, see Davis (1959).

Shivering for thermoregulatory purposes is distinct from muscle tremor. It involves a rhythmic oscillation of skeletal muscle with a frequency that generally is inversely related to body size (Günther, Brunner, and Klussmann, 1983). Whereas cold-induced shivering in mouse and rat has a frequency of 40 and 31 Hz, respectively, the tremor frequency after administration of the chemical tremorine is the same in both species. Although it has not been well studied, it appears that shivering is a more efficient means of producing heat than is muscle tremor. For example, rats administered the toxic chemical chlordecone (Kepone) display overt tremor of the skeletal muscles, but nonetheless sustain a significant decrease in body temperature in spite of a normal metabolic rate (Cook et al., 1987).

Shivering has been considered to be the primary source of facultative heat production in the rat and other rodents acclimated to thermoneutral temperatures. Pharmacological agents that selectively blocking shivering and nonshivering thermogenesis provide an excellent means of dissecting the relative contributions of these thermoregulatory effectors. Griggio (1982) showed that the metabolic rate in the rat acclimated to 30°C (i.e., thermoneutrality) and exposed to 10°C was marginally reduced when nonshivering thermogenesis was blocked with propranolol (a β-receptor antagonist); however, administration of mephenesin, an agent that blocks shivering, led to a marked decrease in metabolic rate during cold exposure (Table 4.2). At thermoneutrality, mephenesin also had no significant effect on metabolic rate. It can also be seen in the cold-acclimated rat that when nonshivering thermogenesis is blocked, it nonetheless exhibits a normal increase in metabolic rate via an increase in shivering. These data clearly show the reliance on shivering as a primary source on thermogenesis in the rat acclimated to thermoneutral temperatures. Moreover, though cold acclimation is associated with a predominant reliance on nonshivering thermogenesis (see Chapter 7), shivering is still available as a thermoregulatory effector during cold exposure.

Most laboratory rodents are maintained at ambient temperatures of 20–24°C, which is below their metabolic TNZs (see Chapter 3). Under these conditions it is likely that most rodent species will utilize both shivering and

Table 4.2. *Metabolic responses of cold-acclimated (10°C) rats when exposed to ambient temperatures of 10°C and 30°C after administration of propranolol and mephenesin, pharmacological blockers of nonshivering and shivering thermogenesis, respectively*

	Metabolic Rate (ml min^{-1} kg$^{-0.75}$)	
Treatment	Cold-acclimated	Warm-acclimated
No drug		
$T_a = 10°C$	32.2 ± 1.3	22.3 ± 0.8
$T_a = 30°C$	16.4 ± 0.9	14.2 ± 0.5
Mephenesin		
$T_a = 10°C$	29.4 ± 0.8	12.0 ± 0.5
$T_a = 30°C$	16.3 ± 0.4	14.3 ± 0.7
Propranolol		
$T_a = 10°C$	26.4 ± 1.4	23.3 ± 1.3
$T_a = 30°C$	15.1 ± 1.0	15.7 ± 0.4

Source: Data from Griggio (1982).

nonshivering mechanisms to thermoregulate. For example, a febrile body temperature in the rat is achieved through activation of shivering and non-shivering thermogenesis (Horwitz and Hanes, 1976; Jepson et al., 1988; Fyda, Cooper, and Veale, 1991a). Nonshivering thermogenesis can be suppressed, while shivering is augmented, in the cold-exposed rat by artificially warming the hypothalamus (Fuller, Horwitz, and Horowitz, 1975). The cold resistance of mice adapted to 25°C is gravely affected by preadministration of propranolol, suggesting that shivering is not the primary heat source for thermoregulation (Estler and Ammon, 1969). On the other hand, thermoregulatory responses in the golden hamster during cold exposure are not markedly affected by administration of propranolol (Vybíral and Janský, 1974). Overall, there would appear to be species-specific differences in the ability to increase shivering when subjected to total β-receptor blockade.

Data on the threshold temperatures for initiation of shivering are surprisingly meager in the rodent literature (Table 4.3). Threshold ambient temperatures for shivering are well below species' LCTs (see Chapter 3). The control of shivering in the cold-exposed rat appears to be heavily influenced by the activity of thermal receptors in the tail. That is, the magnitude of shivering is highly correlated with a decrease in tail skin temperature, but poorly correlated with core temperature and body surface temperature other than in the tail (Davis, 1959). Unfortunately, there are few data on threshold skin temperatures for initiation of shivering in rodents. It would be interesting

Table 4.3. *Comparison of threshold ambient temperatures for initiation of shivering in laboratory rodents*

Species	Threshold T_a (°C)	References
Mouse	20	Oufara et al. (1987)
Rat	15–20	Stoner (1971)
Rat	20 (24% increase in EMG reading compared with $T_a = 30°C$)	Hart et al. (1956)
Rat[a]	23.5	Dawson and Malcolm (1981)
Hamster	15	Pohl (1965)

[a]Partially restrained.

to know which body sites in rodents with and without tails are most important in the activation of shivering.

Problems can arise in ascertaining the threshold temperature (ambient or core) for initiation of shivering: (1) the initiation of shivering may be so subtle as to be indistinguishable from the background muscular activity, and (2) assuming certain similarities between rodents and larger mammals, shivering would appear to be initiated in specific muscle groups, with other muscles being "recruited" into shivering as cold exposure progresses. Such responses would make it difficult indeed to establish clear thresholds for shivering thermogenesis.

4.2.2. Nonshivering thermogenesis

The physiology of nonshivering thermogenesis clearly has been one of the most intensively studied aspects of temperature regulation; for reviews, see Himms-Hagen (1984, 1986, 1990a,b), Trayhurn and Nicholls (1986), and Horwitz (1979). The term *nonshivering thermogenesis* refers to the heat produced from metabolic processes that do not involve the contraction of skeletal muscle (IUPS, 1987). Although by definition it can include the heat produced from basal metabolism, it is customary to define nonshivering thermogenesis as the metabolic heat produced above the basal level in response to cold exposure.

Nonshivering thermogenesis is crucial in three major facets of rodent thermal physiology: (1) it provides the heat needed for thermal homeostasis for many neonatal rodents that lack insulation and have little capacity for shivering at birth (see Chapter 6), (2) it almost completely replaces shivering as the major source of metabolic heat production during cold acclimation (see

Table 4.2 and Chapter 7), and (3) it acts as an energy buffer during overeating, reducing metabolic efficiency and thus serving as a factor in regulating weight gain.

4.2.2.1. Brown adipose tissue. The crux of nonshivering thermogenesis in rodents is the function of the brown adipose tissue (BAT). BAT serves a variety of thermoregulatory functions in rodents: (1) it is a major source of heat during cold acclimation; (2) it provides a substantial proportion of the heat needed to elevate body temperature during fever (Dascombe et al., 1989), as discussed earlier; (3) it serves as a primary heat source during recovery from hibernation, torpor (Lyman, 1990), and anesthetic-induced hypothermia (Shimuzu and Saito, 1991); (4) it appears to be an important site for the activation of diet-induced thermogenesis, as discussed later.

BAT is a truly unique structure. Its main purpose appears to be to produce heat for thermoregulation. It will be recalled that other thermoregulatory effectors have evolved from physiological systems used for nonthermoregulatory functions (Chapter 1). BAT is a useful preparation for many research disciplines because it has a unique biochemical structure and contains a distinct neural innervation, and its function is quite amenable to environmental influences. Central control of BAT appears to be mediated from hypothalamic nuclei, particularly the ventromedial and lateral hypothalamic areas (Himms-Hagen, 1990a,b). The interscapular BAT receives noradrenergic input from the intercostal branches of T1–T4 in the sympathetic nervous system. The firing rate of these efferent neurons can be markedly enhanced in the rat by applying cold stimuli to the face, ear, and neck (Kurosawa, 1991). Neural activity to BAT increases as the ear skin temperature falls below 28°C; however, the firing rate also paradoxically increases as skin temperature rises above 40°C. BAT is also innervated by sensory neurons containing substance P. The functioning of both afferent and efferent neurons appears essential for the normal growth and control of BAT (Himms-Hagen, Cui, and Sigurdson, 1990). Administering capsaicin, the pungent component of chili peppers that depletes substance-P levels in sensory neurons, leads to severe BAT dysfunction (Cui, Zaror-Behrens, and Himms-Hagen, 1990). Moreover, sectioning the sympathetic nerves innervating the interscapular BAT impairs its normal time course of development during cold acclimation (Park and Himms-Hagen, 1988).

Much has been learned over the past 20 years regarding the molecular, cellular, and systemic physiological characteristics of BAT thermogenesis. BAT is distinguished from white adipose tissue (WAT) by several morphological and physiological characteristics. BAT is multilocular (consisting of multiple

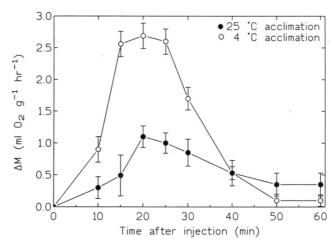

Figure 4.4. Example of increased thermogenic responses to intraperitoneal NE (0.4 mg kg^{-1}) in a normal mouse and a cold-acclimated mouse. Metabolic rate calculated as change from saline controls. Data from Doi and Kuroshima (1982a).

fat droplets), is richly vascularized with dense sympathetic nerve endings, and is densely packed with mitochondria. WAT is unilocular, is poorly vascularized, has relatively few mitochondria, and is affected only by increased levels of circulating hormones, insulin, and epinephrine. Whereas WAT is found throughout most tissues and organs, BAT is found in distinct, strategic locations: interscapular, cervical, pericardial, intercostal, and perirenal regions; for review, see Nechad (1986) and Himms-Hagen (1990a,b). BAT's arterial and venous circulation is structured to facilitate heat transfer from BAT to strategic thoracocervical areas, including the spinal cord and heart (Smith and Roberts, 1964). The interscapular region is by far the most thoroughly studied of the BAT sites. Specific criteria for differentiation of BAT and WAT have been published (Daniel and Derry, 1969; Flaim, Horowitz, and Horwitz, 1976).

BAT is one of the most thermogenic tissues, a function that is augmented in the cold-acclimated state. For example, under proper catecholamine stimulation, BAT is capable of generating heat at the incredible rate of 400 W kg^{-1}, or 80 times the BMR of a rat (Girardier, 1983). In vivo, a rodent's thermogenic capacity can be detected by an elevation in BAT temperature (Flaim et al., 1976) and/or by an increase in metabolic rate following injection of NE or other adrenergic agonists. Indeed, mice acclimated to 4°C have metabolic responses to NE administration that are 340% of those for animals acclimated to 25°C (Figure 4.4). This augmented sensitivity to NE is

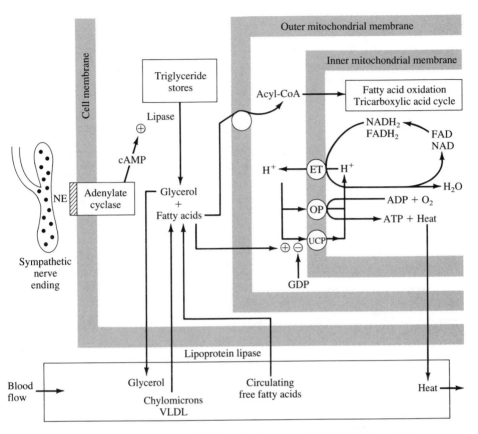

Figure 4.5. Basic cellular mechanism for thermogenesis in BAT. This model was adapted from Power (1989) and Himms-Hagen (1990a, 1990b): ET, electron transport system; OP, oxidative phosphorylation; UCP, uncoupling protein. See text for details.

largely attributed to the development of BAT in the cold-acclimated rodent.

The tremendous thermogenic capacity of BAT is attributable to unique adaptations at the molecular and cellular levels that have been studied in great detail over the past two decades (Himms-Hagen, 1990a,b) (Figure 4.5). Briefly, the inner mitochondrial membrane contains a unique protein referred to as thermogenin, P 32000 polypeptide, or uncoupling protein (UCP, the preferred term), whose function is critical in BAT's thermogenic function. UCP is a proton-translocator that is stimulated by free fatty acids (FFAs) and inhibited by purine nucleotides, especially guanosine diphosphate (GDP). When UCP is operative, the electrochemical proton gradient of the inner mitochondrial membrane collapses, causing reduced efficiency (i.e., uncoupling) in the electron transport chain. The net result is a reduction in the amount of adenosine triphosphate (ATP) synthesized per milliliter of oxygen consumed. Assuming that ATP must always be regulated at a constant level in BAT, the activation of UCP in effect "forces" BAT to use more oxygen and produce greater quantities of heat.

Increased synthesis of messenger RNA (mRNA) for UCP is detected in BAT nuclei within 15 min after acute cold exposure and reaches peak levels by 6–24 hr (Ricquier, 1989; Himms-Hagen, 1990a,b). FFAs serve both as an intermediate signal and as fuel for BAT thermogenesis. The release of NE from BAT sympathetic neurons activates α_1- and β-adrenergic receptors, stimulating lipase and concomitant increased levels of FFAs. BAT has a low capacity for synthesis of ATP and a high capacity for oxidation in the electron transport chain, further enhancing its thermogenic capacity. BAT also contains a form of thyroxine 5′-deiodinase (T5′D) unique compared with that found in other tissues such as liver and kidney (Himms-Hagen, 1990a,b). T5′D produces triiodothyronine (T_3) during NE stimulation. Saturation of BAT nuclear receptors with T_3 enhances the synthesis of UCP, adding further to BAT's thermogenic capacity.

BAT exhibits a variety of physiological and structural changes during cold acclimation, all of which contribute to increase its capacity to generate heat (Foster, 1984; Himms-Hagen, 1986). These changes, which occur to varying degrees in all laboratory rodents, are seen at the subcellular, cellular, and organ levels of BAT and include the following:

↑ UCP levels
↑ GDP binding
↑ mitochondrial cytochrome oxidase activity
↑ mitochondrial density
↑ FFA oxidation enzymes (i.e., peroxisomes)
↑ total BAT protein
↑ T5′D
↑ lipoprotein lipase
↑ metabolic sensitivity to sympathomimetics (e.g., NE)
↑ blood flow during cold exposure and/or sympathomimetics
BAT hypertrophy/hyperplasia

One or more of these variables may be used as indices of increased BAT development (Foster and Frydman, 1979; Wickler et al., 1984; Milner and Trayhurn, 1990). It should be noted that measuring only one or two of these variables may not always provide a complete picture of BAT function. Most commonly, GDP binding, UCP levels, and sensitivity to NE are mainstays in assessing BAT function.

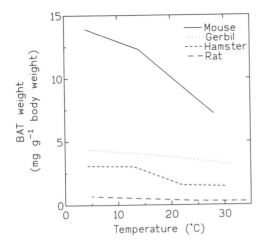

Figure 4.6. Increased mass of interscapular BAT as a function of decreasing acclimation temperature in various laboratory rodents. Data from Hogan and Himms-Hagen (1980) (mouse), Trayhurn and Douglas (1984) (gerbil), Trayhurn et al. (1983) (hamster), and Kuroshima et al. (1982) (rat). Data for mouse include scapular BAT and interscapular BAT.

A principal factor affecting BAT function is body mass. The capacity for nonshivering thermogenesis is generally found to be inversely related to body mass (Chaffee and Roberts, 1971; Janský, 1973). This is most likely attributable to the allometric impact of increased heat loss with decreasing body mass (see Chapter 3). The relative weight of BAT at thermoneutrality, as well as following cold acclimation, is generally greater in species of small body weight (Pospisilova and Janský, 1976) (Figure 4.6). Indeed, the relative weight of BAT in the mouse at thermoneutrality is more than double that in the rat following several weeks of cold acclimation.

In addition to the gross morphological differences in BAT between species, there are quantitative differences at the intracellular level. Development of BAT thermogenesis in the golden hamster during cold acclimation is not accompanied by increased mitochondrial content of UCP as occurs in rat and mouse (Himms-Hagen, 1984). It seems that UCP levels are maintained relatively high regardless of acclimation temperature in the hamster (Nicholls and Locke, 1984). It is also interesting to note that whereas BAT hypertrophy and hyperplasia during cold acclimation occur in all laboratory rodents, there are interspecies differences in metabolic responses to NE. The metabolic response to administration of NE is augmented tremendously during cold acclimation in the mouse and rat, whereas the gerbil and golden hamster show little difference in BAT thermogenesis as a function of temperature acclimation (Vybíral and Janský, 1974; Luebbert, McGregor, and Roberts, 1979). Further details on the function of BAT during cold acclimation are presented in Chapter 7.

4.2.2.2. Diet-induced thermogenesis. In addition to serving as a thermoregulatory effector for facultative metabolism, BAT's highly developed thermogenic capacity allows it to act as an energy "buffer" to excessive caloric intake. The responses of BAT and other tissues to regulate energy balance through changes in heat production are referred to as diet-induced thermogenesis (DIT). DIT is distinct from specific dynamic action, or the heat production associated with the cost of digestion, absorption, and assimilation of nutrients (Stock, 1989). Numerous studies over the past decade have shown a clear relationship between enhanced BAT thermogenesis and excessive food consumption; for reviews, see Girardier (1983), Himms-Hagen (1990a), Trayhurn (1989), and Rothwell, Stock, and Stribling, (1990). Moreover, the development of obesity in rodents is, in many instances, associated with impaired BAT thermogenesis (see Chapter 8). Although BAT function is a focal point of DIT, it should be noted that the liver is also considered to have a major role in DIT (Ma and Foster, 1989).

The metabolic origin of DIT appears to be related to the control of cold-induced thermogenesis. Basically, stimulating DIT in rats by feeding a highly palatable diet (often called a "cafeteria" diet) causes changes in BAT that mimic those of cold acclimation (Rothwell and Stock, 1980) (Figure 4.7). Some of the more notable changes in rodents overfed with a cafeteria diet include an elevated resting metabolic rate, an enhanced metabolic response to NE, reduced shivering during acute cold exposure, increased levels of plasma T_3, increased levels of UCP, and increased weight of BAT (Tulp, Gregory, and Danforth, 1982; Trayhurn, 1989). On the other hand, limiting food consumption or imposing fasting attenuates BAT thermogenic capacity.

An obvious question arises from these data: Does DIT play a role in the normal process of cold acclimation? It is well known that rodents increase their food intake during cold acclimation and acclimatization, and in the wild they increase their food intake in anticipation of cold exposure during the winter (see Chapter 7). It follows that if overeating augments the BAT thermogenic function via activation of the sympathoadrenal system axis, then increased food consumption could represent an important behavior for enhancing cold resistance during the winter (e.g., Tulp et al., 1982). However, research findings regarding this proposition are equivocal. For example,

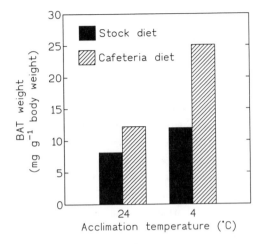

Figure 4.7. Influence of increased caloric consumption on development of BAT thermogenic activity in rat. A palatable "cafeteria" diet increases the weight of BAT, especially following cold acclimation. Data from Rothwell and Stock (1980).

Kuroshima and Yahata (1985) compared thermoregulatory capacities in cold-acclimated rats permitted ad libitum feeding and cold-acclimated, pair-fed controls eating the same amount as warm-acclimated (25°C) animals. It was found that key nonshivering thermogenic characteristics, such as BAT weight, ability to thermoregulate during acute cold exposure, and metabolic response to NE, were generally similar regardless of the quantity of food consumed. On the other hand, Johnson et al. (1982) reported that restricted food intake attenuated BAT weight during cold exposure. Further study of variables such as diet composition, length of feeding, and species specificity should lead to a better understanding of the role of DIT and the development of cold resistance.

4.3. Evaporative heat loss

As ambient temperature increases, the temperature gradients between the skin and body core and between the skin and air diminish. This limits the effectiveness of the transfer of heat from the core to the skin and the rate of dry heat loss from the body to the surrounding environment. It follows that as dry heat loss is reduced with rising ambient temperature, evaporation becomes the only means of heat loss if thermal homeostasis is to be maintained.

The latent heat of varporization of water (λ), defined as the amount of heat absorbed (or released) by evaporation (or condensation), is inversely dependent on the temperature of the water (T) according to the relationship (IUPS, 1987)

$$\lambda = 2,490.0 - 2.34T \tag{4.1}$$

Thus, at 37°C the evaporation of 1.0 g of water absorbs 2,403 J of heat. The dimensions of evaporative water loss (EWL) usually are normalized with respect to body mass (e.g., mg H_2O min^{-1} kg^{-1}). Assuming a constant value of λ (usually 2,403 J g^{-1}), then evaporative heat loss (EHL) can be calculated in dimensions normalized with respect to surface area or body mass (i.e., W m^{-2} or W kg^{-1}). Because these units are equivalent to metabolic rate, this conversion can be useful when partitioning the heat losses under various environmental conditions, such as estimating the proportion between EHL and metabolic heat production, as discussed later.

There are two general routes for EWL in mammals: (1) passive loss of water by diffusion through the skin and that lost through respiration [also referred to as insensible water loss, which is now considered an outdated term (IUPS, 1987)] and (2) active loss of water via sweating, panting, and application of saliva, urine, and other forms of moisture to the fur and skin.

4.3.1. Passive water loss

Although eccrine sweat glands innervated by cholinergic neurons can be identified histologically in the skin of the rat and mouse and can be activated with cholinergic stimulation, they apparently are nonfunctional in thermoregulation in rodents (Hayashi and Nakagawa, 1963; Sivadjian, 1975; Kennedy, Sakuta, and Quick, 1984). However, there is a considerable rate of passive (i.e., insensible) loss of water from cutaneous and respiratory surfaces that contributes significantly to the overall rate of heat loss (Table 4.4).

Studies in rats and other rodents have found that approximately 50% of the passive water loss occurs via diffusion across the skin (Tennent, 1946). For the golden hamster, cutaneous water loss accounts for 70% of the total passive EWL at an ambient temperature of 30°C (Rodland and Hainsworth, 1974). Cutaneous water loss is influenced by the rate of skin blood flow and thus is likely to be affected by changes in temperature (Harlow, 1987). However, relative to the rodent's total heat loss, typical changes in cutaneous EWL are not expected to markedly alter total heat loss.

Respiratory EWL (E_{res}) is highly affected by changes in body temperature and ambient temperature. E_{res} is commonly measured in terms of whole-body

Table 4.4. *Percentage of total heat loss due to EWL in laboratory rodents as a function of ambient temperature*

Species	T_a (°C)	Percentage of total heat loss	References
Mouse	15	7	Hošek and Chlumecký (1967)
Mouse	23	14–18	Hošek and Chlumecký (1967)
Mouse	34	40	Hošek and Chlumecký (1967)
Mouse	20	17	Gordon (1985)
Mouse	31.5	25	Gordon (1985)
Mouse	34.5	36	Gordon (1985)
Hamster	14	6	Gordon et al. (1986)
Hamster	20	7	Gordon et al. (1986)
Hamster	30	10	Gordon et al. (1986)
Hamster	34	17	Gordon et al. (1986)
Hamster	22	4	Jones et al. (1976)
Hamster	40	100	Rodland and Hainsworth (1974)
Rat	15	10	Schmidek et al. (1983)
Rat	20	12	Schmidek et al. (1983)
Rat	23–35	11–16	Tennent (1946)
Rat	30	26	Schmidek et al. (1983)
Rat	35	36	Schmidek et al. (1983)
Rat	36	80–85	Hainsworth (1968)
Rat	40	90	Hainsworth (1968)
Guinea pig	16	11	Gordon (1986)
Guinea pig	22	11	Gordon (1986)
Guinea pig	30	18	Gordon (1986)
Guinea pig	36	28	Gordon (1986)

heat loss and is expressed in dimensions of watts per square meter. Although rodents are not classified as panters, they nonetheless exhibit marked increases in breathing rates when subjected to heat stress (Table 4.5). Panting (i.e., thermal tachypnea), defined as a rapid respiratory frequency accompanied by an increased minute volume and a decreased tidal volume, is activated by the thermoregulatory system to dissipate heat (IUPS, 1987). On the other hand, the increased respiratory frequency of rodents during heating apparently is not a true thermoregulatory response, but rather an indirect effect on respiratory processes as a consequence of increased body temperature. In addition to respiratory frequency, E_{res} is affected by two key environmental variables: the moisture content of the air (i.e., relative humidity) and the ambient temperature. The interaction effects of these environmental factors on respiratory heat exchange have been studied in some rodent species (Collins, Pilkington, and Schmidt-Nielsen, 1971; Welch, 1984).

Table 4.5. *Effect of ambient temperature on respiratory frequency in laboratory rodents*

Species	T_a (°C)	f (breaths min^{-1})	References
Mouse	10	141	Gordon and Long (1984)
Mouse	20	123	Gordon and Long (1984)
Mouse	30	106	Gordon and Long (1984)
Hamster	10	76	Gordon and Long (1984)
Hamster	20	73	Gordon and Long (1984)
Hamster	30	71	Gordon and Long (1984)
Hamster	10–15	77	Reigle and Wolfe (1974)
Hamster	23–28	79	Reigle and Wolfe (1974)
Hamster	33–38	80	Reigle and Wolfe (1974)
Rat	8	90	Lin and Chai (1974)
Rat	25	130	Lin and Chai (1974)
Rat	36	168	Lin and Chai (1974)
Rat	30	123	Nattie and Melton (1979)
Rat	35	188	Nattie and Melton (1979)
Rat[a]	normothermic	84	Richards (1968)
Rat[a]	hyperthermic	152	Richards (1968)
Guinea pig	normothermic	48	Richards (1968)
Guinea pig	hyperthermic	168	Richards (1968)

[a]Anesthetized.

An increase in breathing rate was found to elevate E_{res} in restrained rats, accounting for as much as 21% of the total heat loss at an ambient temperature of 31°C (Lin et al., 1979). On the other hand, Collins et al. (1987) found E_{res} to be a relatively constant ~8 W m^{-2} over a range of ambient temperatures of 8–30°C. E_{res} is more than doubled in hamsters exposed to a temperature of 40°C for at least 20 min (Rodland and Hainsworth, 1974). In spite of these responses, the contribution of passive EWL as a route of heat loss becomes minor as active EWL is recruited at temperatures above thermoneutrality, as discussed next.

4.3.2. Active evaporation

Spreading saliva on the fur and other surfaces is a crucial thermolytic response in rodents subjected to heat stress. This behavior allows rodents to dissipate most of their metabolic heat through evaporation when subjected to ambient temperatures that equal or exceed their body temperatures (Table 4.4). The saliva is applied to bare, vascularized surfaces such as the paws, the scrotum, and the base of the tail, as well as to the fur (Hainsworth, 1967; Stricker and Hainsworth, 1971; Hubbard, Matthew, and Francesconi, 1982;

Matthew et al., 1986). Grooming behavior can be considered analogous to sweating, in that exposed surfaces are covered with moisture, which, during evaporation, increases the level of heat loss and thereby prevents lethal elevations in body temperature. Rats, and presumably other rodents that have been surgically desalivated, are exceedingly susceptible to heat stress (Stricker and Hainsworth, 1971). For example, at an ambient temperature of 40°C, rats whose tails have been amputated have an 8% reduction in heat tolerance, as judged by the time it takes them to reach a core temperature of 40°C. On the other hand, desalivated animals have an 81% reduction in heat tolerance. Furthermore, like sweating in primates, salivation in heat-stressed rats can be blocked by administration of muscarinic antagonists that block sweating in primates, such as atropine (Matthew et al., 1986). Thus, the heat-stressed rat may be a viable experimental model for the study of heat-related illnesses in humans.

The elevation of EWL above passive losses in rodents represents a unique integration of behavioral and autonomic neural processes. That is, an increase in saliva secretion from the salivary glands via parasympathetic stimulation must occur concurrently with grooming behavior in order to maintain an elevated rate of EWL. Thus, thermally activated EWL offers an excellent paradigm for studying the interactions of behavioral and autonomic reflexes. For example, Yanase et al. (1991) have recently reported that the threshold rectal temperature for grooming and salivary flow is 38.2°C in the free-moving rat; below that threshold, grooming and salivary flow coincide with each other, whereas above that threshold there is a copious flow of saliva with no correlation to the incidence of grooming. This suggests that the neural controls over grooming and salivary flow are independent. The independence of these functions is further suggested by the distribution of thermal-sensitive sites in the rat CNS for their initiation; salivation seems to predominate with heating of the POAH area, whereas grooming is activated only when the posterior hypothalamus is heated (Figure 4.8). Although these processes appear to be independent, obviously there must be some level of coordination in order to achieve a steady level of EWL during heat stress.

Normal grooming behavior accounts for 7–8% of total EWL in rats maintained at an ambient temperature of 25°C (MacFarlane and Epstein, 1981). It is interesting to note that during water deprivation, a rat will continue to groom in spite of the fact that such behavior increases its water loss. Thus, it appears that grooming behavior occurs independently of those reflexes involved in water conservation. This is indeed a peculiar independence of control, because heat stress and water deprivation often occur simultaneously under natural environmental conditions. The time a rat spends grooming with

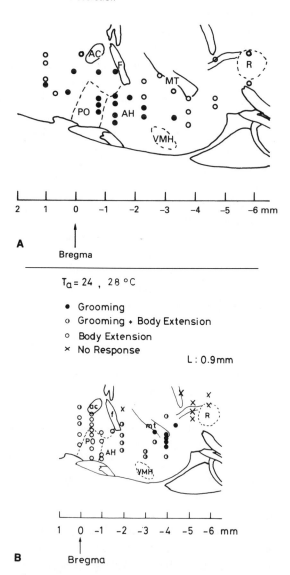

Figure 4.8. Localization of thermal-sensitive sites in the brain stem of the rat that elicit salivation (A) and grooming and body extension (B). The POAH specifically elicits salivation, whereas the posterior hypothalamic area elicits grooming during localized heat stimulation. Reprinted from Yanase et al. (1991) with permission from *The Journal of Physiology* (A) and from Tanaka et al. (1986) with permission from Pergamon Press (B).

saliva increases dramatically as temperature is increased above the TNZ (Hainsworth, 1967). As heat stress continues, the ensuing dehydration eventually limits active EWL; lethal hyperthermia is imminent when the percentage of water loss from the plasma exceeds 15% (Hainsworth, Stricker, and Epstein, 1968).

Grooming behavior is quite adaptable to unusual environmental circumstances. For example, when mice and rats are exposed to heat stress (40°C) and have free access to a pool of water, rather than drink they thermoregulate behaviorally by grooming the water onto their fur, and they survive these conditions for up to 8 hr, compared with 3–6-hr survival when no water is available (Stricker, Everett, and Porter, 1968). Moreover, rats surgically desalivated and exposed to heat stress will groom with urine as a means of increasing EWL; however, compared with rats that have normal salivary function, this response is relatively ineffective for heat dissipation (Hubbard et al., 1982; Matthew et al., 1986).

In general, active EWL is initiated in rodents as the ambient temperature is increased above the TNZ. Just as the demarcation between the basal metabolic rate and a cold-induced increase in metabolic rate is difficult to discern (see Chapter 3), the ambient temperature at which active EWL is initiated can be as difficult or more difficult to identify. Threshold temperatures for active EWL vary in rodent species exposed to relatively dry air conditions: ~34°C for mice (Gordon, 1985), 30–36°C for various strains of rats (Gordon, 1987), and 36°C for guinea pigs (Gordon, 1986). In anesthetized rats, the threshold core temperature for salivation is 3.0°C higher than that needed to induce vasodilation in the tail (Nakayama et al., 1986); hence, the salivation/grooming response is most likely the last autonomic effector to be recruited for heat dissipation during heat stress.

The partitioning of EWL into the cutaneous, respiratory, and salivary components changes dramatically as ambient temperature is elevated above the TNZ (Table 4.6). When exposed to acute heat stress (40°C), the rat and hamster initially exhibit similar peak elevations in EWL; however, as exposure continues, the EWL of the hamster falls relatively rapidly, whereas that of the rat is maintained for several more hours before heat exhaustion occurs (Figure 4.9). The seeming lack of data on EWL in the guinea pig is unfortunate. Lacking a tail for heat dissipation, this species would be expected to have a well-developed ability for EWL. Future comparisons between the guinea pig and other rodents should prove interesting.

The submaxillary salivary gland in the rat appears to be the most important source of water for EWL during heat stress (Elmer and Ohlin, 1971; Hainsworth and Stricker, 1971; Horowitz, Argov, and Mizrahi, 1983).

Table 4.6. *Partitioning of EWL into salivary, cutaneous, and respiratory avenues in rat and hamster exposed to ambient temperatures of 30°C and 40°C*

Species	T_a (°C)	Percentage of total EWL		
		Salivary	Cutaneous	Respiratory
Rat	30	0	77	23
Rat	40	60	13	27
Hamster	30	0	73	27
Hamster	40	70	8	22

Source: Data from Rodland and Hainsworth (1974).

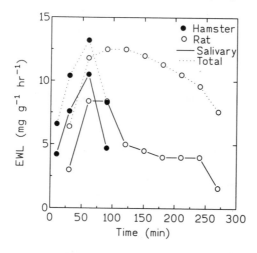

Figure 4.9. Time courses of total EWL and salivary EWL in hamster and rat while subjected to acute heat stress (T_a = 40°C). Total EWL in the hamster is closely related to salivary EWL. Total EWL in the rat is maintained at an elevated level, while salivary EWL decreases with continued heat exposure. Data from Hainsworth et al. (1968) and Rodland and Hainsworth (1974).

Threshold core temperatures for increasing salivation vary from 38°C to 39.5°C and probably are dependent on the prevailing ambient temperature and other factors (Horowitz et al., 1983; Horowitz and Meiri, 1985; Yanase et al., 1991). At an ambient temperature of 36°C the submaxillary gland secretes saliva, while the parotid gland is quiescent; at 40°C the salivary secretion from the submaxillary gland is three times that from the parotid gland (Stricker and Hainsworth, 1971). During heat stress, salivation is increased primarily through activation of parasympathetic cholinergic neurons (Elmer and Ohlin, 1971; Rodland and Hainsworth, 1974; Hubbard et al., 1982).

However, it was recently found that the parotid glands in rats exhibit an increase in NE content and increased density of β-receptors within 24 hr of heat exposure, suggesting that sympathetic outflow must also be considered in the heat adaptation of the salivary glands (Fujinami et al., 1991).

In summary, one can view the relationship between salivary-gland function and heat exposure to be analogous to that of BAT function and cold exposure, in that there are marked biochemical changes in these organs during thermal stress; however, knowledge of the processes in the latter clearly is ahead of that in the former. Future cellular and molecular studies of the salivary gland in heat-stressed rodents should prove an enlightening endeavor in adaptational physiology.

4.4. Behavioral thermoregulatory effectors

Among all the homeostatic processes, it can be argued that the thermoregulatory system is unique in its reliance on behavior as a predominant means of achieving regulation. The behavioral sensing of temperature is exquisitely sensitive, and behavioral effectors for modulation of the ambient thermal environment can be shown to operate continuously. Changes in skin and core temperatures provide nearly continuous input to the thermoregulatory control centers, with subsequent corrective motor responses (see Chapter 1).

Behavioral thermoregulatory effectors can be grouped into two major categories: (1) *natural,* which I define here as the inherent behaviors displayed without requiring any specialized apparatus, and (2) *instrumental,* which are the behaviors observed only with the use of specialized laboratory apparatus (e.g., thermoclines or operant systems). Instrumental responses generally are easier to quantify and are by far the most thoroughly studied of the two forms of behavior (Weiss and Laties, 1961; Satinoff and Hendersen, 1977; Gordon, 1990a). On the other hand, natural behaviors, though not easy to quantitate, are nonetheless critical effectors for thermal homeostasis in rodents.

4.4.1. Natural behavior

Natural thermoregulatory behaviors can be readily observed in rodents exposed to heat and cold stress. Some natural behaviors are integrated with the activation of autonomic thermoregulatory effectors, such as grooming coupled with saliva secretion, as discussed earlier. In the cold, such behaviors can be as simple as an individual animal huddling to restrict heat loss in a cold environment, or as complex as hoarding food (Fantino and Cabanac, 1984), changing the quantity, frequency, and duration of feeding (Johnson and Cabanac, 1982), and socially coordinating the efforts of a group of animals to huddle and increase the insulation of a nest (Batchelder et al., 1983). Thermotaxis is clearly a natural thermoregulatory behavior; however, it is normally measured using elaborate devices, and it is covered later as an instrumental response.

Social aggregation in the cold also lessens the impact of cold stress on caloric requirements, metabolic expenditure, and BAT development. For example, at 25°C, food consumption per unit body mass is reduced by 15% when mice are housed in groups of five, as compared with single animals; at −3°C, food consumption is reduced by 28% in grouped animals, as compared with individuals (Prychodko, 1958). At an ambient temperature of 10°C, pairs of mice allowed to huddle in a nest have 18% less BAT than do single mice maintained at the same temperature (Heldmaier, 1975). It is pertinent to note from these data that standard housing temperatures for rodents can have an impact on food requirements and BAT development, depending on whether the animals are maintained individually or in groups. Clearly, such information is crucial to a variety of disciplines in the life sciences, particularly the fields of nutrition and metabolism.

When subjected to extreme heat stress, rodents will display several natural behaviors designed to enhance heat loss. In a comparative study of several laboratory rodents, Roberts, Mooney, and Martin (1974) measured the latency to activate several natural behaviors, such as grooming of saliva, extension of the body, and placing the head on the floor (Figure 4.10). Extension of the body maximizes the surface-area : body-mass ratio, thereby enhancing heat loss (see Chapter 3). Body extension is achieved with minimal metabolic expenditure, thus further lessening the heat load. The latencies for both body extension and placing the head on the floor seem to be sensitive indications of dire heat stress in all species. The gerbil and hamster appear to be most sensitive in the display of these behaviors. The latency to groom saliva is not clearly affected by temperature, which perhaps is attributable to the ubiquity of grooming behavior in rodents regardless of the environmental conditions.

Rodents placed in warm environments usually exhibit increased motor activity (Clark, 1971; Roberts et al., 1974; Roberts, 1988). Increasing the temperature from 26.6°C to 41.4°C elicits marked increases in motor activity in the hamster, gerbil, rat, and guinea pig, whereas the mouse remains relatively inactive (Roberts et al., 1974). This hyperactivity in the other species probably can be attributed to the animals' search for an escape from the stressful environment and increased grooming of saliva onto the fur. The survival of rats subjected to heat stress is inversely dependent on spontaneous locomotor

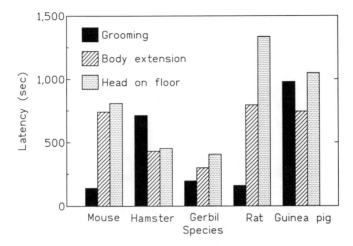

Figure 4.10. Comparative survey of natural thermoregulatory behaviors in rat, guinea pig, gerbil, hamster, and mouse when subjected to heat stress (T_a = 37.7°C). Behavior assessed in terms of latency for grooming, body extension, and placing head on floor within a 45-min period after beginning of heat exposure. Data from Roberts et al. (1974).

activity (Clark, 1971). Discounting escape, an increase in motor activity is detrimental in the long run because of the increased metabolic heat load.

Thiessen and colleagues, working with the Mongolian gerbil, have found unique interactions between a simple behavioral effector, grooming, and the secretory function of the harderian gland (Thiessen et al., 1977; Thiessen and Kittrell, 1980; Pendergrass and Thiessen, 1983; Thiessen, 1988; Grant and Thiessen, 1989). The harderian gland, located in the orbit of the eye, becomes activated in the cold and secretes a complex mixture of lipids and porphyrins. As the gerbil grooms, the lipid material from the gland is spread on the fur and increases the insulation of the fur. Also, because the gland secretions darken the fur, the gerbil's absorbance of solar radiation is enhanced. Hence, the gland's secretions serve to limit heat loss as well as augment heat gain in animals maintained in a cold environment.

In the heat the gerbil will exhibit increased sand-bathing behavior, which aids in removal of the harderian lipids from the fur, thereby lightening the color of the coat, increasing its solar reflectance, and reducing the insulative quality of the fur (Pendergrass and Thiessen, 1983). Its grooming behavior, coupled with increased harderian secretion in cold environments and increased salivary secretion and sand bathing in hot environments, constitutes a rather complex interaction of autonomic and behavioral thermoregulatory effectors for this species.

4.4.2. Instrumental behavior

Instrumental thermoregulatory behavior in rodents has been assessed using two principal devices: the temperature gradient (or thermocline) and operant systems that permit selection of warm or cold reinforcements. Each method has its advantages and drawbacks (e.g., Laughter and Blatteis, 1985; Gordon, 1990a): The conditions of a temperature gradient are most likely to be conducive for display of inherent behavior, whereas operant systems involve extensive training; usually there is easy access to the test subject in an operant system, whereas temperature gradients can be cumbersome and do not offer easy access to the animal; operant systems provide a precise quantitative measure of a given thermoregulatory behavior (e.g., number of thermal reinforcements per unit of time), whereas temperature gradients provide a gross measure of thermal selection; in an operant system the animal must continually perform a motor task in order to prevent an incipient increase or decrease in ambient temperature, whereas in a temperature gradient the selected T_a can be achieved quickly with no further muscular effort. This latter point can be important in studies where one wishes to study thermoregulatory behavior during a sleep–wake cycle and in other situations where experimental manipulation impairs normal motor activity and could thereby affect thermoregulatory behavior when tested in an operant system.

4.4.2.1. Behavior in temperature gradients. When placed in a continuum of air and/or surface temperatures, animals exhibit thermotropic behavior, a turning or a movement in response to a thermal stimulus, and seek out a *thermopreferendum,* a condition that presumably represents a preferred environment for heat exchange (IUPS, 1987). Thermotropic behavior is the prevalent effector in the lower vertebrates (reptiles, amphibians, and fish) as well as invertebrates and undoubtedly was the first thermoregulatory effector to develop in the evolution of homeothermy (Whittow, 1970; Prosser, 1973). Thermotropic behavior is a favored effector because it uses so little metabolic energy. A variety of instrumental studies have shown that rodents preferentially utilize behavioral rather than autonomic effectors to thermoregulate during heat and cold stress (Gordon, 1983b; Schmidt, 1984; Morimoto et al., 1986).

The term ''thermopreferendum'' can refer to the preferred temperature of the skin, core, air, or substrate. However, because core temperature in rodents

Table 4.7. *Selected T_a values for sexually mature laboratory rodents placed in a temperature gradient*

Species	Selected T_a	References
Mouse	30.1	Eedy and Ogilvie (1970)
Mouse	31.8	Ogilvie and Stinson (1966)
Mouse	29.5–31.9	Gordon (1985)
Gerbil	32.4	Eedy and Ogilvie (1970)
Gerbil	28.2	Thiessen et al. (1977)
Gerbil	32.3	Akins and Thiessen (1990)
Hamster	30.2	Gordon et al. (1984)
Hamster	28.2	Gordon et al. (1986)
Rat	27.2	Marques et al. (1984)
Rat	24.6	Spencer et al. (1990)
Rat	19.8–24.9	Gordon (1987)
Rat	31	Refinetti and Carlisle (1986b)
Rat	30–31	Ettenberg and Carlisle (1985)
Rat[a]	19	Refinetti and Horvath (1989)
Rat[a]	21–24[b]	Corbit (1970)
Guinea pig	29–30	Blatteis and Smith (1980)
Guinea pig	30.6	Gordon (1986)

[a] Animals tested in operant-type apparatus.
[b] Within this range of T_a values it was predicted that there was no operant selection for cool air or radiant heat.

is normally stable over a wide range of ambient temperatures, the thermopreferendum is generally expressed in terms of the preferred or selected T_a. Although these terms are essentially synonymous, preferred T_a may be considered by some to be anthropomorphic, whereas selected T_a more accurately describes a measure of the animal's behavior.

Selected T_a values for common laboratory rodents have been determined by various laboratories (Table 4.7). In most cases the selected T_a coincides closely with a species' metabolic TNZ (see Chapter 3). That is, the species selects a T_a associated with minimal metabolic expenditure. The guinea pig is unusual in that it selects T_a values near the upper end of the TNZ, whereas the hamster, mouse, and gerbil select T_a values near the middle or lower end of the TNZ (Figure 4.11).

The rat exhibits an altogether distinct behavioral response when placed in a temperature gradient. First, as can be seen from the data of Table 4.7, there is wide variability in selected T_a values from various laboratories. Many studies have found that the selected T_a of the rat is well below its TNZ. Poole and Stephenson (1977a) first noted that the rat's metabolic TNZ was markedly

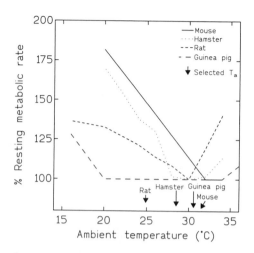

Figure 4.11. Relationships between selected T_a and metabolic profile for hamster, rat (Sprague-Dawley), golden hamster, and mouse. Animals tested for periods of approximately 2 hr. Note relatively cool selected T_a for rat. Data from Gordon (1985, 1986, 1987) and Gordon et al. (1986).

higher than the ambient temperature zone for normal behavioral activity. Rats given free range in a temperature gradient select cool temperatures and have higher metabolic rates than do rats restricted from selecting temperatures no lower than the LCT (Gordon, 1988). Several studies have concluded that the laboratory rat selects relatively cool temperatures associated with a metabolic rate above its basal level, thus contradicting the general premise that animals thermoregulate behaviorally to minimize energy expenditure. However, in further study of this issue, we recently found that a naive rat placed in a temperature gradient will gradually select warmer ambient temperatures, reaching its zone of metabolic thermoneutrality up to 6 hr after placement in the gradient (Gordon et al., 1991). It is likely that the rat's initial preference for cool temperatures is attributable to stress due to exposure to a novel environment, but further work on this issue is needed. Overall, the rat shows a distinct temporal response in a temperature gradient compared with other rodents. In spite of its popular use, the rat may not be the most suitable model of behavioral thermoregulation.

There can be a complex interaction between behavioral thermoregulation and social behaviors in rodents. For example, when a dominant gerbil and a subordinate gerbil are placed together in a gradient, the dominant individual selects the normal T_a of 28°C, whereas the subordinate individual is relegated to a cooler T_a of 22°C (Thiessen et al., 1977). When the odor from mice subjected to the stress of electric shock to the foot is directed into a temperature

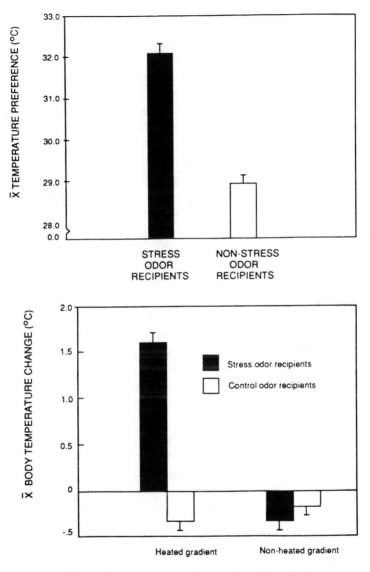

Figure 4.12. Effects of odor from stressed mice on selected T_a and body temperature of unstressed mice in a temperature gradient. Reprinted from Thiessen et al. (1991) with permission from Plenum Publishing Corp.

gradient containing a nonstressed mouse, it selects warmer ambient temperature and exhibits an elevation in body temperature (Figure 4.12). Odors from stressed animals are valuable cues used to indicate distress in a population. This novel thermoregulatory response appears to be part of an anticipatory physiological response that facilitates an organism's resistance to potential environmental stresses (Thiessen, Akins, and Zalaquett, 1991).

4.4.2.2. Behavior in operant systems.   The majority of the operant thermoregulatory studies have been carried out in rats. There have been very few studies in other rodent species that would allow for appropriate comparisons. Operant systems basically utilize a paradigm in which the animal is motivated to avoid a thermally adverse cold or hot environment by operating a lever to receive a reward of heat or cold, respectively (Weiss and Laties, 1961; Carlisle, 1968; Satinoff and Hendersen, 1977). In hot environments, rats can be trained to actuate a reward system consisting of cool air or a shower to accelerate heat loss (Epstein and Milestone, 1968; Szymusiak et al., 1985). In a cold environment, mice (Baldwin, 1968) and rats (Weiss and Laties, 1961; Carlisle, 1968; Satinoff and Rutstein, 1970; Refinetti and Carlisle, 1986a) will operate an infrared lamp for heat rewards. Refinetti and Horvath (1989) developed a system allowing a rat to actuate a cold-air intake or a warm-air intake and thereby regulate its ambient temperature. Interestingly, the rats (Long-Evans strain) selected a temperature of 19°C, which is similar to that observed in short-term measurements of selected T_a for this strain in a temperature gradient (Table 4.7). The range of temperatures where operant thermoregulatory responses are absent would implicate this environment as a zone of behavioral thermoneutrality. Using such logic, Corbit (1970) estimated a behavioral TNZ of 21–24°C. It is interesting to note that, as in several temperature-gradient studies, this range of temperatures is below the rat's metabolic TNZ.

Operant systems provide an excellent means of assessing the interaction between behavioral and autonomic thermoregulatory processes. Because the operant behavior is easier to quantitate, changes in environmental stimuli are rapidly reflected in the animal's behavioral responses. For example, in rats trained to bar-press for various intensities of radiant heat at an ambient temperature of −5°C, the hypothalamic temperature is maintained at a constant level even though the intensity of the heat reward is increased by fivefold (Carlisle, 1968). This homeostasis is achieved because of the animal's ability to reduce the frequency of its heat rewards and thereby maintain a steady level of heat intake in the cold. Refinetti and Carlisle (1986a) further studied the complementary nature of operant thermoregulation by measuring metabolic heat production and behavioral heat intake simultaneously using a scenario in which the force needed to depress the reward lever was increased (Figure 4.13). When the force requirement was increased, the rat opted to increase its heat production rather than expend the additional energy needed

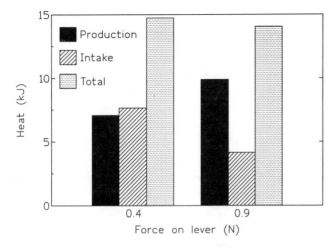

Figure 4.13. Example of complementary roles of behavioral thermoregulation and autonomic thermoregulation in a rat exposed to an ambient temperature of 3°C and trained to bar-press for radiant heat. Data from Refinetti and Carlisle (1986a).

to maintain the same rate of behavioral heat intake. The trade-off between metabolic rate and behavioral heat intake led to an approximate balance in total heat intake, resulting in a constant core temperature.

4.5. Motor activity as a thermoregulatory effector?

Early studies noted that spontaneous motor activity, generally measured as wheel-running, increased in rats during periods of food deprivation (Campbell and Lynch, 1967, 1968) and cold exposure (Fregly, 1956; Finger, 1976; Poole and Stephenson, 1977a). It was thought that the increase in motor activity was a thermoregulatory-mediated response to raise heat production and thus counteract the decreased metabolic rate caused by food deprivation and the increased heat loss due to cold exposure; for review, see Gordon (1990a). Indeed, a common misconception among novices is that because of their small sizes and high rates of heat loss, rodents must remain active to maintain a normal body temperature.

Some studies of rats have suggested that an increase in motor activity is not a true thermoregulatory effector. For example, when food-deprived rats are maintained at a warm ambient temperature of 31°C to prevent a decrease in body temperature, they are nonetheless more active than controls fed ad libitum (Campbell and Lynch, 1967). Apparently the increased motor activity during food deprivation is attributable mostly to food-searching behavior. It should also be added that whereas motor activity in the cold increases heat production, it also accelerates heat loss because of increased convection. A cold-exposed endotherm will thermoregulate more efficiently by huddling to prevent heat loss.

The reasons behind the increased motor activity as a function of decreasing ambient temperature remain elusive. This is an important issue, because motor activity in rodents is a key variable for a variety of fields, such as psychobiology and behavioral pharmacology. To the best of my knowledge, no experiment has completely ruled out motor activity as a thermoregulatory effector in cold-exposed rodents. Obviously, heat is produced during motor activity and is used to maintain thermal homeostasis. It has been shown that warm- and cold-acclimated rats substitute the exercise-induced heat load for the heat normally generated by shivering and nonshivering thermogenesis (Hart and Janský, 1963; Arnold et al., 1986) (see Section 3.5). But it is also clear that inactive rats will nonetheless regulate their core temperatures at approximately the same level as active animals. One point to consider is that the rise in body temperature during running activity at warm ambient temperatures serves as a cue for cessation of activity, thus preventing overheating (Campbell and Lynch, 1967; Gordon and Heath, 1980). With this in mind, rather than accepting the proposal that cooler temperatures increase motor activity, it could be hypothesized that rodents have an internal drive to be active (especially at night) and that this natural increase in activity is limited by elevations in ambient temperature. Clearly, the issue of motor activity as an effector during cold exposure needs further study.

5

Body temperature

Body temperature is the regulated variable of the thermoregulatory system. That is, the thermoreceptors, integrative neural systems, and thermoregulatory motor outputs function to maintain thermal homeostasis throughout the body. Measuring the body temperatures of normal and compromised individuals can provide insight into the nature of a species' thermoregulatory mechanisms. This chapter attempts to explore the properties of rodents' body temperatures, including variations in normal core temperature, ambient limits of normothermia, modulations in temperature from day to night and during sleep, and the responses to psychological stress.

5.1. Partitioning of body temperature

Thermal physiologists conventionally divide body temperature into three components: (1) the core, including the body trunk (e.g., spinal cord, abdominal and thoracic cavities); (2) the shell, which includes the skin and subcutaneous tissues that are directly affected by changes in ambient temperature; (3) the brain. The internal or core body temperature is perhaps the most frequently measured parameter in thermal physiology. The thermal core is composed of "inner tissues of the body whose temperatures are not changed in their relationship to each other by circulatory adjustments and changes in heat dissipation to the environment" (IUPS, 1987). Thus, the temperature of any organ or tissue in the thermal core is considered to be extremely stable over a wide range of ambient temperatures. Body sites such as the brain, spinal cord, rectum, colon, esophagus, and visceral and thoracic organs usually are considered to be parts of the thermal core. However, as will be shown later, the temperatures at some of these sites occasionally are independent of one another and thus cannot always be said to represent a common core temperature.

5.1.1. Mean body temperature

Researchers not familiar with the latest definitions in thermal physiology often use the term "body temperature" as a measure of core temperature. However, "body temperature" can be an ambiguous term, because, in literal definition, it can refer to the temperature at any site in the body. Because there is almost always a heterogeneous distribution of temperatures in a homeotherm, a given measure of body temperature will be spatially unique.

Mean body temperature $(\overline{T}_b)$, which is an integration of every temperature in the body, is an ideal measure for the thermal physiologist. It is equal to the sum of the products of the heat capacities $(c_i m_i)$ and temperatures (T_i) of all tissues of the body divided by the total heat capacity of the organism (IUPS, 1987):

$$\overline{T}_b = \Sigma(c_i m_i T_i)/\Sigma(c_i m_i) \tag{5.1}$$

where c is specific heat, and m is mass.

Clearly, it is not practical to measure the temperature and specific heat at every site in the body. However, if the core temperature (T_c) and mean skin temperature $(\overline{T}_{sk})$ are known, then mean $\overline{T}_b$ can be calculated:

$$\overline{T}_b = a_1 T_c + a_2 \overline{T}_{sk}, \quad \text{with } a_1 + a_2 = 1 \tag{5.2}$$

Factors a_1 and a_2 represent the contributions of the thermal core and the shell to mean body temperature, respectively. The ratio $a_1 : a_2$ varies from approximately 9 : 1 to 6 : 4, depending on environmental temperature and other factors (IUPS, 1987). For example, exposure to a relatively cool ambient temperature results in peripheral vasoconstriction, a subsequent increase in the thickness of the peripheral shell, and hence a higher value for a_2. It should be noted that the concept of equation (5.2) is based on studies in human subjects; little is known about the values of a_1 and a_2 in rodents.

Mean body temperature can also be measured accurately in animals using direct calorimetry (Hart, 1951). In this method, a rodent equilibrated to a given environment is quickly killed and placed in a water-filled, insulated flask. With proper calibration, the heat content of the animal can be determined from the increase in water temperature. Dividing the animal's heat content (joules per gram of body weight) by its specific heat (the whole-body average usually is 3.47 J g^{-1} $°C^{-1}$), yields the mean temperature. Under most conditions the thermal shell is cooler than the core, and mean body temperature is consistently below that of the core temperature. For example, in mice the difference between mean body temperature and core temperature increases as ambient temperature is reduced from 30°C to 0°C (Hart, 1951).

Although it is rarely measured in rodent studies, the parameter of mean body temperature can nonetheless provide useful information on a species' thermoregulatory characteristics. Knowledge of mean body temperature can be very useful in the calculation of heat storage (see Chapter 1).

5.2. Core temperature

The core temperature can be determined using a variety of methods: (1) insertion of a thermocouple or thermistor probe into the rectum or colon; (2) surgical attachment of probes to various anatomical sites such as the abdomen, thorax, or brain; (3) implantation of biotelemetry devices that transmit temperature information from awake, unrestrained animals to a remote radio receiver. Use of a colonic probe is the simplest, least expensive method and is used most frequently. The depth of penetration of the probe is crucial for accurate estimation of core temperature. Varying the depth of insertion has shown that the measured core temperature falls dramatically at distances of less than 5 cm from the anal opening for rat (Lomax, 1966), and 7 cm for guinea pig (Czaja and Butera, 1986). For mouse, gerbil, and hamster, the insertion distance for accurate measurement of core temperature is approximately 2.5 cm.

Core-temperature data collected with colonic probe and implanted thermocouples are generally similar (Table 5.1). Although it has not been studied in much detail, there is some evidence of thermal heterogeneity in the core. For example, at an ambient temperature of 28°C the temperatures of the liver, mesentery, and lower abdomen in a conscious rat were found to be 39.3°C, 39.1°C, and 38.5°C, respectively (Grayson and Mendel, 1956). Oxygen consumption in the liver is higher than that in its immediate surroundings, especially during ingestion of a meal. A rat's liver temperature can be as much as 0.5°C above that of the blood in the portal vein (Adachi, Funahashi, and Ohga, 1991). There can be marked differences in temperatures in the colon and esophagus during the onset of and recovery from hypothermia in the rat and hamster (Adolph and Richmond, 1955). Activation of BAT thermogenesis by cold exposure or pharmacological stimulation creates "thermal pockets" in the interscapular region and areas of the core receiving blood from BAT sites (see Chapter 4). There is also a marked thermal heterogeneity between the trunk and scrotum in the male rat, as discussed in more detail in Chapter 6.

Measurements of core temperatures using a variety of methods indicate that the mouse, hamster, gerbil, and rat maintain average daytime core temperatures of 36.5–38.0°C, whereas the guinea pig exhibits a slightly warmer

Table 5.1. *Core body temperatures* (T_c) *of normothermic laboratory rodents measured under resting, unrestrained conditions during daylight hours*[a]

Species	T_a (°C)	T_c (°C)	References
Mouse (cln) [a]	27	36.4	Herrington (1940)
Mouse (cln)	18–31	36.9–37.3	Gordon (1985)
Mouse (cln)	21–23	37.1	Muraki and Kato (1987)
Mouse (cln)	21	36.0–36.5	Connolly and Lynch (1981)
Mouse (abd)	25	36.3–37.6	Kluger et al. (1990)
Gerbil (cln)	23–30	38.1–38.6	Robinson (1959)
Gerbil (cln)	10–32	37.9–38.2	Luebbert et al. (1979)
Gerbil (cln)	25–30	37.9–38.0	Klir et al. (1990)
Hamster (tc)	30	38.2	Pohl (1965)
Hamster (cln)	22	36.8	Jones et al. (1976)
Hamster (cln)	14–32	36.0–36.3	Gordon et al. (1986)
Hamster (abd)	23	37.3–37.8	Conn et al. (1990)
Hamster (es)	~25	37.1–37.4	Gordon and Fogelson (1991b)
Hamster (hy)	~22	36.8	Duncan et al. (1990)
Rat (cln)	25	36.6	Herrington (1940)
Rat (cln)	20–24	37.0–37.9	Gordon (1987)
Rat (abd)	23	37.1	Spencer et al. (1976)
Rat (abd)	25	37.2	Thornhill et al. (1978)
Guinea pig (cln)	21	38.7	Herrington (1940)
Guinea pig (cln)	22	38.4	Werner (1988)
Guinea pig (cln)	20–22	38.6	Czaja and Butera (1986)
Guinea pig (cln)	22–30	38.1–38.3	Gordon (1986)
Guinea pig (abd)	25	38.6	Thorne et al. (1987)

[a]Parentheses indicate sites of measurement: cln, colon or rectum; abd, abdominal cavity; tc, thoracic cavity; es, esophagus or stomach; hy, hypothalamus.

core temperature of 38.0–39.2°C (Table 5.1). It is important to recognize that these are average temperatures, and the distribution of core temperatures in a large population can vary by more than 1.0°C from the mean (Figure 5.1). In this example, core temperatures varied by as much as 1.6°C between two strains of rats under identical housing conditions. It would be of interest to determine if that variation was simply biological "noise" or was an indication of individual differences in set-point.

The psychological stress from handling and insertion of colonic temperature probes has marked effects on core temperatures in laboratory rodents. The stress due to insertion of a colonic probe can lead to a 1.0°C rise in temperature, and the animal may not fully recover for 3 hr (Poole and Stephenson, 1977b; Lotz and Michaelson, 1978; Gallaher et al., 1985). It should be emphasized that the initial insertion of a colonic probe gives an accurate reading of core temperature if it is measured within 5–10 sec – the stress-

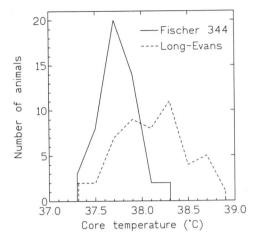

Figure 5.1. Frequency distributions of core (colonic) temperatures for Fischer 344 albino and Long-Evans pigmented rat strains. Temperatures were measured in animals maintained in their home cages in standard animal housing facilities maintained at 22°C and 50% relative humidity. Data from Gordon and Fogelson (1991a).

induced change in core temperature takes several minutes to develop. Simple handling and the presence of personnel in the animal quarters can also lead to significant elevations in core temperature (Georgiev, 1978). Thus, repeated measurements or chronic attachment of a colonic temperature probe in re-strained animals are likely to give artificially elevated readings of core temperatures in rodents (see Section 5.6).

5.2.1. Effects of ambient temperature on core temperature

As discussed in Chapter 4, thermoregulatory motor outputs for heat dissipa-tion in a warm environment and heat gain/conservation in a cold environment function to regulate core temperature. The limits of normothermia, defined here as the range of ambient temperatures above and below which thermal homeostasis is overwhelmed, provide important information on a species' thermoregulatory characteristics. Unfortunately, there are relatively few data on the limits of homeothermy in rodents (Table 5.2). Interspecies compari-sons of the ambient limits of homeothermy can be difficult because the ex-posure time at a given temperature is a crucial factor that affects the upper and lower limits of normothermia.

The demarcation between normothermia and hyperthermia is easier to dis-cern than the break between normothermia and hypothermia. Generally,

Table 5.2. *Threshold upper and lower ambient temperatures for maintenance of normothermic core temperatures in laboratory rodents*

Species	Upper limit (°C)	Lower limit (°C)	References
Mouse	32.5	~15	Herrington (1940)
Mouse	32	—[a]	Gordon (1985)
Mouse	34	5	Oufara et al. (1987)
Gerbil	32	<4	Luebbert et al. (1979)
Gerbil	35	<−10	Klir et al. (1990)
Hamster		<−15	Pohl (1965)
Hamster	32		Gordon et al. (1986)
Rat	28	13	Herrington (1940)
Rat	28–34	—	Poole and Stephenson (1977a); Gordon (1987)
Rat		10	Krog et al. (1955)
Rat		<−5.7	Depocas et al. (1957)
Guinea pig	~26.5	—	Herrington (1940)
Guinea pig	30	20	Gordon (1986)

[a]Not Reported

slight elevations in core temperature occur as ambient temperature is equal to or just exceeds the upper critical temperature. On the other hand, unrestrained laboratory rodents are well adapted to defend against a decrease in core temperature when exposed to relatively cold temperatures. Because of the rodent's well-developed cold resistance, the criteria for the lower ambient temperature of normothermia can be ambiguous (Lovegrove et al., 1991). Not surprisingly, a smaller species such as the mouse is least capable of thermoregulating in the cold, whereas the gerbil, hamster, and rat have been shown by some investigators to remain normothermic at temperatures below 0°C. Considerable variation has been reported for the lower limit of normothermia in the rat, which probably can be attributed to differences in time of exposure. For example, Herrington (1940) noted a decrease in colonic temperature in rats maintained at 14–15°C for 6 hr, whereas Depocas et al. (1957) found that rats remained normothermic when exposed to −5.7°C for 30–135 min. The reported limits of normothermia in the guinea pig are equivocal. A recent study found the upper and lower limits to be 30°C and 20°C, respectively (Gordon, 1986). On the other hand, 6 hr of exposure to a temperature of 16.5°C led to a slight elevation in core temperature in the guinea pig (Herrington, 1940).

The data in Table 5.2 define the relatively broad ambient limits of normothermia in rodents. It is interesting to note that within this range of temperatures are subtle changes in core temperature that, though not well studied, can nonetheless reveal details of the neural characteristics of the species' thermoregulatory system. For example, over a range of ~15–35°C, core temperatures in the rat and guinea pig exhibit a slight U-shaped function, with the lowest core temperature occurring at ambient temperatures of 22–32°C (Herrington, 1940; Gordon, 1987). The rise in core temperature with increasing ambient temperature probably is a result of subtle body warming that is not compensated by heat-dissipating responses. On the other hand, the rise in core temperature with decreasing ambient temperature appears to be linked to an overcompensation of nonshivering thermogenic mechanisms. Morimoto et al. (1986) found that core temperature in the rat increased by 1.0°C when ambient temperature was lowered from 23°C to 0°C. This was accompanied by an abrupt rise in temperature in the interscapular BAT that could be blocked by administration of a β-blocker (propranolol). The rise in core temperature during ambient cooling was associated with a marked increase in metabolic activity in the ventromedial hypothalamus, a site implicated in the control of BAT thermogenesis (see Chapter 4). Moreover, when rats were trained to thermoregulate behaviorally by bar-pressing for radiant heat, their core temperatures remained unchanged during cold exposure. Thus, the rise in core temperature during cold exposure may be a result of overcompensating autonomic effectors.

5.3. Brain thermal homeostasis

Conduction of action potentials, membrane transport, synaptic transmission, and other processes crucial to the functioning of the CNS are exceedingly temperature-dependent (Janssen, 1992). Clearly, brain thermal homeostasis is essential in the face of environmental heat and cold stress, fever, and exercise. In neuropharmacological studies, the purported effects of chemical agents on the CNS sometimes can be attributed to a change in brain temperature. Moreover, as body mass decreases, the influence of ambient temperature on brain temperature in rodents becomes more marked (Tegowska, 1991). Thus, a thorough understanding of temperature control in the brain in rodents is essential to many fields of study in the life sciences.

The brain had conventionally been considered to be part of the thermal core. It was generally assumed that brain temperature was either equal to or paralleled changes in the temperature at more accessible core sites, such as the colon, esophagus, or abdomen. However, research over the past few decades has shown that many avian and mammalian species are capable of selective brain cooling during exercise and/or heat stress. Presumably, this response is needed to protect the brain from overheating; for review, see

Baker (1982). Brain cooling was thought to be operative only in those species possessing a carotid rete, including cat, various ungulates, and other species. The carotid rete is a specialized vascular structure in which warm blood from the trunk passes via the internal carotid arteries through a network of fine blood vessels (a rete) that is bathed in relatively cool venous blood draining into the cavernous sinus. The cooled arterial blood is then directed to the brain. Rodents, rabbits, and primates lack a carotid rete; hence, little attention was given to studying selective brain cooling in these species. However, it is now recognized that varying degrees of brain cooling occur in species without retes, including humans (Cabanac, 1986).

Although the responses are not as marked as in animals possessing carotid retes, it is clear that rodents exhibit an independence of brain temperature and trunk temperature. The brain temperature of the rat can vary by as much as 1.0°C during the nocturnal cycle (Abrams and Hammel, 1965). An inactive rat that begins to feed will exhibit an increase in brain temperature and a decrease in rectal temperature (Rampone and Shirasu, 1964). A guinea pig exposed to a high-pressure helium-oxygen atmosphere, which results in hypothermia, is capable of maintaining its brain temperature above rectal temperature and thereby retaining neural function during hypothermia (Unger, Hempel, and Kaufmann, 1980). One study of mice has shown that brain and rectal temperatures are closely correlated with each other at ambient temperatures of 16–34°C, with brain temperature consistently below colonic temperature (Benjamin et al., 1987). It was further shown that brain temperature in ethanol-intoxicated mice increased less rapidly than did colonic temperature with rising ambient temperature, suggesting a mechanism for independent control of brain temperature over trunk temperature in this species. In another mouse study, brain temperature was slightly above rectal temperature during ambient cooling, but when the mice were exposed to heat stress (T_a = 40°C), brain temperature was 0.2°C below rectal temperature (E. Sedunova, personal communication, 1992).

There have been many studies showing an independence of brain and trunk temperatures in rodents; however, studies on the mechanisms of its physiological control are meager. Thermal homeostasis in the brain, as in any tissue or organ, is dependent on a balance between heat production and heat loss. On the heat-production side, variations in brain temperature can be correlated with shifts in arterial blood flow from the trunk (Abrams et al., 1965). There is also evidence that the rat brain exhibits an increase in metabolism during ambient cooling, thereby providing heat for the control of brain temperature (Szelenyi and Donhoffer, 1978). On the heat-loss side, the cranium in rodents is encompassed with venous sinuses, which when filled with relatively cool blood can augment heat loss (Gordon et al., 1981; Caputa, Kadzie, and Narebski, 1983; Caputa, Kamari, and Wachulec, 1991). Blood in these sinuses is cooled either through direct heat exchange with the skin or through evaporative cooling of the nasal mucosa, in which cooled blood is directed into the venous sinuses surrounding the cranium. A relatively simple experiment in anesthetized animals can exemplify the role of the nasal mucosa in brain thermal homeostasis (Gordon et al., 1981): When a hamster's nasal passageways are occluded, forcing the animal to breath through its mouth, there is an abrupt increase in brain temperature (Figure 5.2A). It will be important in future studies to understand how the interaction between ambient temperature and relative humidity affects evaporation from the nasal mucosa and hence the control of brain temperature.

The work of M. Caputa has shown that both rat and guinea pig are capable of selectively cooling the brain below the temperature of the trunk during exposure to high ambient temperatures and exercise (Figure 5.2B). Brain cooling is evident only when the trunk temperature exceeds ~40.5°C; the cooling response of the guinea pig appears to be more vigorous than that of the rat (Caputa, Wasilewska, and Swiecka, 1985; Caputa et al., 1991). Maintenance of the brain temperature below 40.5°C is crucial, because cellular metabolic processes in the brain, but not necessarily other tissues and organs, are drastically altered at temperatures above 40.5°C (Caputa et al., 1983).

We are only beginning to understand how brain temperature can affect various physiological and behavioral processes in rodents. For example, Blumberg, Mennella, and Moltz (1987) found that the copulating male rat exhibits transient cooling of the preoptic area during ejaculation (Figure 5.2C). Interestingly, that brain cooling was not necessarily related to increased motor activity, but seemed to be specifically related to copulatory behavior. Some studies have shown that rising brain temperature may be an important cue leading to cessation of running activity in the guinea pig (Caputa et al., 1985), golden hamster (Gordon and Heath, 1980), and perhaps other species. Changes in brain temperature have also been shown to be involved in the transition between awake and sleeping states and in the occurrence of rapid-eye-movement sleep in the rat (Alfoldi et al., 1990).

A confound of selective brain cooling is that hypothalamic cooling during exercise would be expected to stimulate heat-conserving responses and cause a further elevation in temperature. That finding therefore places into question the role of hypothalamic thermal-sensitive neurons in thermoregulation (see Chapter 2). Such a response apparently is bypassed during exercise. Clearly,

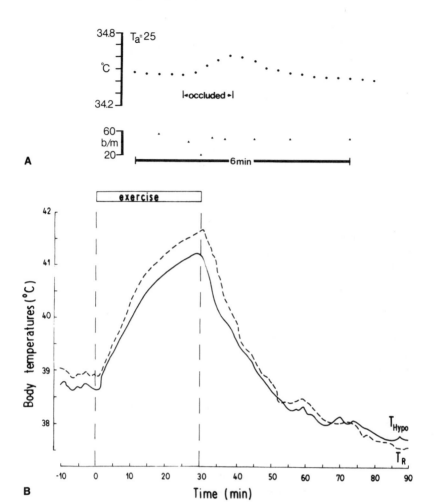

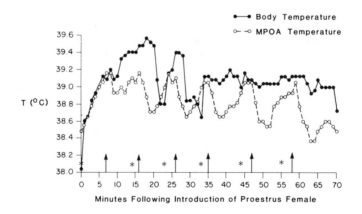

measuring brain and body temperatures in rodents should lead to a better understanding of the role of brain thermal homeostasis in the control of various neurological processes.

5.4. Thermal tolerance

A species' thermal tolerance is defined in terms of survival time and/or lethal core temperature when exposed to acute heat or cold stress. Thermal tolerance is an extremely useful parameter in situations where the total thermoregulatory "fitness" of a species is to be estimated. Measurements of the responses of autonomic and behavioral effectors to thermal stimulation (see Chapter 4) and the ambient limits of normothermia, as discussed earlier, provide considerable data on the sensory, integrative, and motor functions of thermoregulation; however, a species' thermal tolerance is indeed the principal criterion of the capacity of the thermoregulatory system. In other words, whether or not a species survives exposure to a thermally adverse environment is determined solely by the capacity of its thermoregulatory system.

Unfortunately, thermal tolerance currently is not an attractive method of study, because death usually is the biological endpoint. Stricter guidelines for animal care and experimentation implemented over the past decade clearly have had a negative impact on such studies. However, early studies of the thermal tolerance of unrestrained animals have provided a valuable data base on the thermoregulatory capacities of laboratory rodents.

5.4.1. Upper limits of thermal tolerance

When a rodent is exposed to heat stress so acute as to overwhelm its thermoregulatory system, the increase in its core temperature during an

---

Figure 5.2. Examples of selective brain cooling in laboratory rodents. A: Occluding the nasal passageways of an anesthetized hamster leads to an abrupt elevation in brain temperature and a transient decrease in breathing rate. Reprinted from Gordon et al. (1981) with permission from The American Physiological Society. B: Exercising a rat at an ambient temperature of 27°C causes the hypothalamic temperature (T_{Hypo}) to decrease below rectal temperature (T_R) at the point when core temperature exceeds 41°C. Reprinted from Caputa and Kamari (1991) with permission from Pergamon Press. C: Brain cooling during ejaculation in a male rat at an ambient temperature of 21–23°C. Arrows indicate times of ejaculation, and asterisks indicate points of intromission. Reprinted from Blumberg et al. (1987) with permission from Pergamon Press.

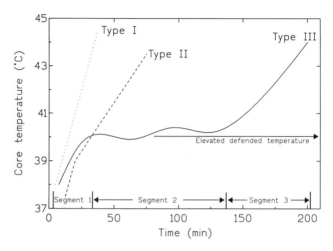

Figure 5.3. Illustration of the type I, II, and III core temperature responses in rodents when exposed to acute heat stress. Type I and II responses are indicative of low thermal tolerance and are common in rodents with impaired thermoregulation. Type III prevails in most rodent species. Data from Erskine and Hutchison (1982a) and Wright et al. (1977a).

experiment will display one of three characteristic functions, conventionally termed types I, II, and III (Figure 5.3) (Erskine and Hutchison, 1982a). The most common response, type III, is composed of three segments: segment 1 is a brief period during which core temperature is elevated by 2–3°C above normal, termed the elevated defended temperature (EDT); segment 2 is a prolonged period during which the EDT is maintained relatively stable but nonetheless increases very slowly; segment 3 is a period during which core temperature sharply increases again as thermoregulation collapses and thermal death is imminent. Ohara, Furuyama, and Isobe (1975) found that key parameters of the type III response, including the EDT and its slope in segment 2, could be used to reliably predict survival time for rats exposed to an ambient temperature of 42.5°C. The type I and II responses are occasionally observed in animals with impaired heat-dissipation mechanisms, and they basically fail to regulate core temperature at an EDT during heat stress (Stricker and Hainsworth, 1971; Wright, Iams, and Knecht, 1977a). The lengths of segments 1, 2, and 3 and the EDT vary between rodent species (Table 5.3). For example, the EDT of the gerbil is 1.0°C below that of the mouse. The rat exhibits the longest segment 2 compared with the mouse and gerbil. The mouse, gerbil, and rat typically display type III responses during acute heat stress.

Table 5.3. *Comparison of various parameters of heat tolerance in the mouse, gerbil, and rat exposed to an ambient temperature of 40°C*

Parameter measured[a]	Mouse	Gerbil	Rat
Initial T_{col}	37.85 ± 0.3[b]	37.83 ± 0.28	37.96 ± 0.19
EDT (°C)	41.05 ± 0.11	39.98 ± 0.17	40.38 ± 0.17
Time in segment 1 (min)	20.39 ± 1.17	16.53 ± 3.49	27.01 ± 1.95
Time in segment 2 (min)	50.61 ± 6.34	118.08 ± 26.36	153.47 ± 20.06

Source: Data from Erskine and Hutchison (1982a).
[a]See Figure 5.3 for explanation of variables.
[b]Data given as means ± S.E.

Thermal tolerance can be assessed around the time of thermoregulatory collapse by measuring the critical thermal maximum (CT_{max}) and/or the lethal temperature (Hutchison, 1980). CT_{max} was originally developed by Cowles and Bogert (1944) in studying thermoregulation in reptiles, and it is defined as "the thermal point at which locomotor activity becomes disorganized and the animal loses its ability to escape from conditions that will promptly lead to its death." That definition has been modified to include the core or brain temperature at which abrupt muscular spasms occur (e.g., Hutchison, 1980; Erskine and Hutchison, 1982a). CT_{max} is lower than the lethal body temperature and essentially represents an ecological maximum. That is, CT_{max} will be lethal only if the organism is incapable of escaping from life-threatening conditions. It is often difficult to clearly define the points of CT_{max} and lethal temperature, because mammals often will survive very high body temperatures, but die several hours later (Adolph, 1947; Erskine and Hutchison, 1982a). Interestingly, mice that survived exposure to CT_{max} and were returned to 25°C became markedly hypothermic for at least 3 hr (Wright, 1976). Such "rebounds" in thermoregulation are not well characterized and deserve further study.

Values for lethal temperature and CT_{max} have been reported for many rodent species (Tables 5.4 and 5.5). The upper limit of thermal survival is affected by many environmental factors, including ambient temperature, relative humidity, water availability, exposure time, degree of restraint, activity level, previous thermal history (e.g., acclimation), and circadian cycle (Adolph, 1947; Hart, 1971). Among laboratory rodents, the guinea pig appears best able to withstand high temperatures, followed by the rat and mouse. Body size clearly is a crucial factor that limits thermal tolerance; with increased size, the rate of rise in core temperature is generally attenuated,

Table 5.4. *Lethal core temperatures* (T_c) *and survival times* (ST) *for laboratory rodents exposed to acute heat stress*

Species	Exposure T_a (°C)	T_c (°C)	ST (min)	References
Mouse	44	44.6	26	Frankel (1959)
Hamster	39–41	42.5–42.8	~75	Kilham and Ferm (1976)
Hamster	50	45.1	27	Frankel (1959)
Rat	42.5	44.77–45.37	68–132	Furuyama (1982)
Rat	42.5	44.1–44.9	~95	Isobe et al. (1980)
Rat	38–50	42.5 (LD$_{50}$)	60–480	Adolph (1947)
Rat	50	44.8	39	Frankel (1959)
Guinea pig	44–50	42.8 (LD$_{50}$)	33–480	Adolph (1947)

Table 5.5. *Critical thermal maximum* (CT_{max}) *and minimum* (CT_{min}) *and survival time for laboratory rodents subjected to acute heat or cold stress*

Species	T_a (°C)	CT_{max} (°C)	Time to CT_{max} (min)	References
Mouse	40	42.25–43.4	56–142	Erskine and Hutchison (1982b)
Mouse	40	42.62	71	Erskine and Hutchison (1982a)
Gerbil	40	44.0	136	Erskine and Hutchison (1982a)
Rat	40	44.22	180	Erskine and Hutchison (1982a)

Species	T_a (°C)	CT_{min} (°C)	Time to CT_{min} (min)	References
Mouse	−40	16–20	19–26	Ferguson and Folk (1970)
Mouse	−18	15.6	30–99	Gordon and Ferguson (1980)
Gerbil	−40	16–20	26–51	Ferguson and Folk (1970)
Hamster	2[a]	16.9[b]	—[c]	Panuska et al. (1969)
Hamster	−40	16–20	39–62	Ferguson and Folk (1970)
Rat	−40	16–20	39–60	Ferguson and Folk (1970)
Guinea pig	2[a]	25.3[b]	—	Panuska et al. (1969)
Guinea pig	−20	15.7[d]	294	Hirvonen et al. (1976)

[a]Fur clipped around trunk and neck.
[b]Indicates temperature of loss of neck-righting reflex.
[c]Not reported.
[d]Represents lethal temperature.

thus maximizing thermal tolerance. The body-size factor may well account for the superior thermal tolerance of the guinea pig compared with smaller rodent species. Thermal tolerance has also been shown to vary as a function of the time of day in the rat, although data from different sources are contradictory. One study found that heat tolerance was maximal at noon and minimal at midnight (Isobe, Takaba, and Ohara, 1980), whereas another study found heat tolerance to be best between 1600 and 2000 hr, and least from 800 to 1200 hr (Wright, Knecht, and Wasserman, 1977b). One of the main factors affecting thermal tolerance in rats is their ability to groom saliva during heat stress (Furuyama, 1982, 1988). Some genetic strains of rats exhibit better saliva-grooming behavior and thus prolong their survival time during acute heat stress. Adolph (1947) clearly demonstrated that increased relative humidity had deleterious effects on thermal tolerance in rodents and other mammals. Increased humidity will limit the effectiveness of evaporation of saliva from the fur and respiratory mucosa, leading to rapid overheating (see Chapter 4). Not surprisingly, thermal tolerance in restrained rats is greatly diminished compared with unrestrained animals (Frankel, 1959).

5.4.1.1. Heat stroke. Extreme elevations in core temperature can lead to heat stroke, which often is associated with permanent pathological sequelae and, under severe conditions, death (e.g., Goodman and Knochel, 1991). Heat stroke usually is defined as an excessive rise in body temperature resulting from an overload or failure of the thermoregulatory system during heat stress. In heat-stressed rodents, the point of heat stroke occurs during the latter stages of type I and II responses as well as during segment 3 of the type III heating (Figure 5.3). Heat stroke is thought to result either from heat-induced damage to the thermoregulatory centers in the CNS or from circulatory failure, resulting in shock and an acute rise in body temperature.

The rat has been used as a model for studying the circulatory and thermoregulatory mechanisms of heat stroke (Hubbard et al., 1978; Kregel, Wall, and Gisolfi, 1988). For example, when a rat is exposed to acute heat stress, there are progressive increases in heart rate and mean arterial blood pressure (MAP). As core temperature passes 41.5°C there is a sudden fall in MAP, along with marked changes in the resistance of some arterial beds. This indicates that the circulatory shock of heat stroke is attributable to selective loss of vasoconstrictor tone in the splanchnic circulation (Kregel et al., 1988; Kregel and Gisolfi, 1990).

Hubbard et al. (1978) found that exercise-induced elevations in body temperatures of rats in cold and warm environments were more deleterious than were elevations in core temperature due simply to exposure to high ambient temperatures (Figure 5.4). In that case, the LD_{50} core temperature for rats exposed to warm temperatures was 42.4°C, whereas for rats exercised and exposed to a warm temperature, the LD_{50} was reduced to 41.8°C. The

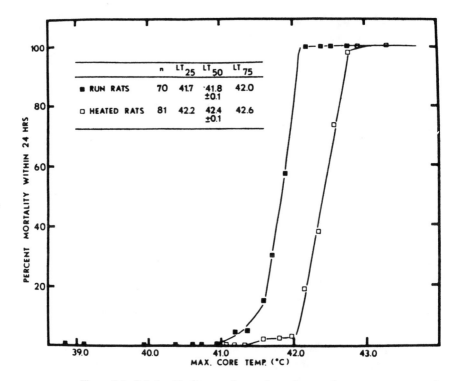

Figure 5.4. Relationship between the maximum increase in core temperature and the percentage mortality for rats either run to exhaustion at a T_a of 5–30°C (filled squares) or immobilized and exposed to a T_a of 41.5°C (open squares). Values in insert give mean ±S.E. core temperatures at 25, 50, and 75% mortality. Reprinted from Hubbard et al. (1978) with permission from The American Physiological Society.

difference in threshold core temperatures would suggest that heat stroke and related illnesses are dependent on factors other than simply high tissue temperature.

5.4.1.2. Adaptability of thermal tolerance.   With the increasing use of local and whole-body hyperthermia for some forms of cancer therapy there is a precedence for understanding the adaptability of thermal tolerance in rodents and other mammals. For example, do repeated hyperthermia treatments alter a species' sensitivity to subsequent hyperthermia treatments? Knowledge of the adaptation of thermal tolerance will be crucial in cancer therapies in which repeated hyperthermia treatment is utilized as an antitumorigenic agent.

There is some evidence regarding the adaptability of thermal tolerance in rodents. A single exposure of a mouse to a core temperature of 42°C causes

Table 5.6. *Example of how thermal tolerance is enhanced in anesthetized rats[a]*

Interval prior to thermal challenge (hr)	Time to reach LD_{50} core temperature (min)
0	25.1
24	55.1
48	56.5
72	50.1
96	50.8
120	33.0
144	22.1

Source: Data from Weshler et al. (1984).
[a]Rats were subjected to a conditioning treatment of 41.8°C core temperature for 60 min. The time to LD_{50} core temperature of 42.5°C was determined at various intervals after the initial conditioning treatment.

an increase in the time to reach its CT_{max} for at least 72 hr after the initial hyperthermia treatment (Wright, 1976). Whereas CT_{max} is unchanged following a single hyperthermia treatment (core temperature = 42°C), CT_{max} is increased by 1.2°C in mice acclimated to 30°C as compared with 15°C-acclimated animals (Wright, 1976). Otherwise, there is little information on the effects of temperature acclimation on thermal tolerance.

A single "conditioning" hyperthermia treatment in an anesthetized rat (core temperature of 41.8°C for 60 min) will lead to a doubling of its thermal tolerance (defined as time to heat death) for up to 120 hr following the conditioning treatment (Table 5.6). This adaptation is not related to enhanced development of thermoregulatory mechanisms, because the tests were performed in anesthetized animals. Moreover, it can be shown that similar increases in heat tolerance occur in specific organs and in vitro preparations subjected to single episodes of acute heating; for review, see Weshler et al. (1984). The study of adaptation to thermal tolerance at the cellular and whole-animal levels should continue to be a fruitful endeavor for understanding the interactions between antitumorigenesis and heat therapy.

5.4.2. Lower limits of thermal tolerance

Cold resistance in rodents is relatively well developed in comparison with heat tolerance. As mentioned earlier, inducing hyperthermia in rodents is a relatively simple procedure whereby the animal is exposed to an ambient temperature of ~5°C above the upper critical temperature for a brief period of time. On the other hand, when exposed to a cold ambient temperature of 0°C (i.e., ~30°C below the lower critical temperature), rodents such as the rat, hamster, and guinea pig may continue to thermoregulate for days before succumbing to hypothermia. Thus, to study cold tolerance over relatively brief periods of time, investigators must have an extremely cold environmental chamber with a T_a well below 0°C. Immersion in an ice-water bath can also be used to rapidly induce hypothermia, but this form of cold stress clearly is not equivalent to exposure to low air temperatures. Alternatively, a novel method to induce hypothermia is exposing animals to an 80 : 20 helium : oxygen atmosphere (Helox) at ambient temperatures of −10°C to 0°C (Musacchia and Jacobs, 1973). Helium has a very low density and a high specific heat compared with nitrogen. Helox has a thermal conductivity four times greater than that of a nitrogen : oxygen mixture. Thus, convective heat loss in a Helox atmosphere is much greater than in normal air (Clarkson, Schatte, and Jordan, 1972). For example, a rat exposed to Helox at an ambient temperature of 0°C will reach a core temperature of 15°C within 145 min (Steffen and Musacchia, 1985).

Measuring the lower limits of thermal tolerance can be hampered by the equivocality of hypothermic death. The core temperature at which breathing and heart contractions cease may vary, making it difficult to confirm cold death. As with heat tolerance, the concept of ecological death can also be useful for studying the lower limits of thermal tolerance. Ferguson and Folk (1970) used the critical-thermal concept and measured the critical thermal minimum (CT_{min}), defined as the core temperature at which the animal could not right itself from a supine position when exposed to acute cold stress ($T_a = -40°C$) (Table 5.5). CT_{min} values were reported to be within the range of 16–20°C for mouse, gerbil, hamster, and rat. As is the case with CT_{max}, CT_{min} is not always lethal. For example, whereas cessation of breathing and loss of righting reflex in the rat occur at a core temperature of 15°C, the heart will continue to beat at temperatures below 10°C provided that artificial respiration is administered (Adolph and Richmond, 1955). The LD_{50} core temperature below which the rat is unable to spontaneously rewarm is 23°C at an ambient temperature of 5°C, but only 15°C for the hamster (Adolph and Richmond, 1955). Thus, given proper environmental conditions for recovery (i.e., a relatively warm ambient temperature), many rodent species can recover spontaneously after their core temperatures have been lowered to CT_{min}.

Although the rat is a nonhibernating species, it can withstand prolonged reductions in core temperature. Using a hypercapnia/hypoxia method to lower temperature, which involves placing the animal in a sealed jar under refrigeration, a rat's core temperature can be lowered to 0–1°C and the animal successfully rewarmed using microwave radiation (Andjus and Lovelock, 1955).

An adult rat can be kept alive for 15 hr at a core temperature of 15°C, but can be revived only after hypothermic durations of less than 5.5 hr (Popovic, 1960). Subjecting rodents to prolonged hypothermia causes marked alternations in carbohydrate metabolism and depletions of glycogen stores. Interestingly, carbohydrate metabolism and ability to recover from hypothermia are significantly improved in the hamster by pretreatment with corticosteroids (Deavers and Musacchia, 1979). On the other hand, the rat's survival to prolonged hypothermia is not enhanced by corticosteroid treatment (Steffen and Musacchia, 1985).

5.5. Circadian temperature rhythm

The core temperature in most endotherms waxes and wanes with a 24-hr periodicity; for review, see Refinetti and Menaker (1992). Along with the circadian temperature rhythm (CTR), other physiological and behavioral processes show clear circadian rhythms, such as metabolism, food and water consumption, heart rate, motor activity, and others. The CTR is an extremely resilient process, but it is susceptible to various types of stress. For example, chronic stress elicited by electric shock to the foot will cause an abrupt cessation of the CTR in the rat for approximately 5 days; however, the CTR will eventually recover in spite of the presence of the stress (Kant et al., 1991). Under conditions of continuous light or darkness (i.e., no zeitgeber for synchronization of physiological rhythms), the CTR of the rat will drift out of phase with real time and, after several months, may show signs of decay in its amplitude (Satinoff, Liran, and Clapman, 1982; Eastman and Rechtschaffen, 1983). The core temperature of the rat has to be reduced to at least 28°C for 6 hr before one sees significant phase delays in wheel-running activity (Gibbs, 1981).

The suprachiasmatic nucleus (SCN) of the hypothalamus is considered to be the 24-hr pacemaker responsible for eliciting circadian rhythms; for review, see Rusak and Zucker (1979). Lesioning the SCN will eliminate the circadian rhythms of many behavioral and physiological systems, but there is somewhat of a controversy regarding the stability of the CTR. One study found that the CTR in rats persisted following SCN lesioning, whereas other rhythms were disrupted (Satinoff and Prosser, 1988). On the other hand, Eastman, Mistlberger, and Rechtschaffen (1984) found that complete SCN lesions eliminated the CTR and 24-hr sleep rhythm of the rat. Moreover, lesioning the medial preoptic area does not affect the period of the CTR, but markedly increases the CTR amplitude (as high as 12°C), which then slowly recovers over a period of several months (Satinoff et al., 1982) (see

Chapter 2). It is clear that the CTR will continue to be a major focus of chronobiology and thermal physiology.

The commonly used laboratory rodents display a breadth of unique CTR characteristics. It is well known that the laboratory rat is nocturnal, being most active during the dark phase of the circadian cycle. The same is true for the hamster and mouse (Refinetti and Menaker, 1992). The gerbil displays most of its spontaneous activity during the night (Roper, 1976); however, under natural conditions this species will switch from diurnal activity patterns in the winter to nocturnal in the summer (Randall and Thiessen, 1980). The circadian responses of the guinea pig are poorly understood. In contrast to the other rodents, there is no clear 24-hr rhythm of metabolism in the guinea pig (Stupfel et al., 1989). Similarly, two studies have found that the core temperature of the guinea pig is extremely stable over a 24-hr period (Baciu et al., 1985; Thorne, Yeske, and Karal, 1987). It would be premature to conclude that the guinea pig lacks a CTR. Yet the possibility that the CTR is either extremely damped or absent from the guinea pig would aid in explaining why this species' daytime core temperature is generally above those of other rodents (see Table 5.1). In other words, comparisons between the daytime core temperatures of nocturnal species and those of the guinea pig (which either has no CTR or is diurnal) are clearly flawed.

Characterizing CTRs using conventional techniques such as periodic insertion of colonic probes or attached thermocouples in restrained animals is not appropriate for rodent studies. The stress resulting from these procedures will induce temperature changes that will confound the natural characteristics of the CTR. On the other hand, the development of computerized biotelemetry systems over the past decade has facilitated CTR research in unrestrained rodents. These systems allow one to monitor core temperature, ECG, EEG, EMG, motor activity, and other crucial variables in unrestrained rodents (Figure 5.5). Certainly, with this technology, breakthroughs in rodent CTR research will be forthcoming in the near future.

The majority of CTR research in rodents has been conducted using rats. The rat's core CTR shows several distinct phases when placed under a 12 : 12 light : dark cycle (lights on at 0600 hr): (1) a stable daytime phase, with a mean core temperature of 37.3°C; (2) a rising phase that begins at 1600 hr, with the steepest increase at 1730–1900 hr; (3) another plateau phase between 1900 and 0500 hr, with a mean core temperature of 38.1°C; (4) a phase during which core temperature drops rapidly and exhibits a transient undershoot of the afternoon temperature (Scales and Kluger, 1987). The CTR in mouse and hamster is generally similar to that in the rat, with the possible exception of interspecies differences in CTR amplitudes (Table 5.7). Hamsters on a 14 : 10

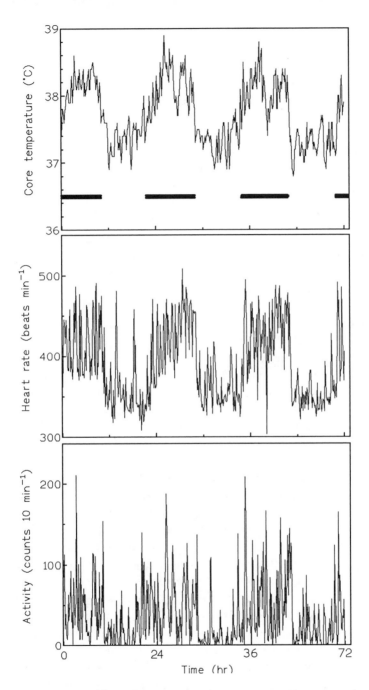

Figure 5.5. Examples of the time courses of core temperature, heart rate, and motor activity in an unanesthetized, unrestrained rat over a 72-hr period. Note the correlation between bursts of motor activity and transient elevations in core temperature and heart rate. Black bars indicate dark periods. Data obtained using surgically implanted telemetry system (Data Sciences). Courtesy of A. H. Rezvani and C. J. Gordon.

Table 5.7. *Survey of characteristics of CTRs in laboratory rodents on a 12 : 12 light : dark photoperiod, with lights on at 0600 hr; body temperature measured by radiotelemetry*

Species	CTR amplitude (°C)	Time of minimum T_c (hr)	Time of maximum T_c (hr)	References
Mouse	1.7	1330	1930	Kluger et al. (1990)
Hamster (sedentary)	1.2	1330	1930	Conn et al. (1990)
Hamster (exercise)	1.4	1330	2130	Conn et al. (1990)
Rat	1.1	0730	2000	Scales and Kluger (1987)

light : dark cycle (a photoperiod that assures normal reproductive status in this species) begin to show a rise in core temperature 2–3 hr before the start of the dark period (Duncan, Gao, and Wehr, 1990). In some cases the hamster's CTR can split into two distinct circadian components under conditions of constant illumination (Pickard, Kahn, and Silver, 1984). In general, the mouse, hamster, and rat on a 12 : 12 light : dark cycle (lights on at ~0600 hr) exhibit relatively stable core temperatures during the daylight hours, a pattern that coincides with the normal working schedule of laboratory personnel. However, it should also be recognized that the body temperatures and physiological and behavioral processes of these species are in their quiescent stages during the daytime.

What are the underlying thermoregulatory mechanisms responsible for generating the CTR in rodents? It was originally thought that motor activity during the nocturnal phase was a leading cause for the increased temperature. In telemetered animals, correlations between small bursts of motor activity and transient elevations in core temperature are clearly evident (Figure 5.5). However, several studies have shown that the nocturnal rise in temperature can be independent of the level of motor activity (Honma and Hiroshige, 1978; Conn et al., 1990). The CTR amplitude for golden hamsters given access to a running wheel is markedly higher during the nighttime than for animals without running wheels; however, hamsters without running wheels nonetheless display a clear nighttime elevation in core temperature (Figure 5.6). Likewise, rats in running wheels have CTR values that range from 36.1°C in the light to 39.0°C in the dark, whereas without a running wheel the CTR ranges from 35.6°C to 38.1°C (Satinoff, Kent, and Hurd, 1991). Thus, it appears that the amplitude of the CTR is internally set by the CNS, but motor activity can nonetheless increase its amplitude during the night.

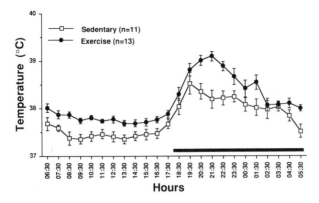

Figure 5.6. Average core body temperatures over a 24-hr period in golden hamsters with and without access to a running wheel. Dark bar indicates period of darkness. Reprinted from Conn et al. (1990) with permission from The American College of Sports Medicine.

5.5.1. Set-point and the CTR

In addition to motor activity, behavioral and autonomic thermoregulatory effectors play major roles in the control of the CTR. During the night, metabolic heat loss from the rat lags behind heat production, resulting in heat storage and hence an elevation in core temperature (Sugano, 1983; Shido, Sugano, and Nagasaka, 1986). Moreover, autonomic heat-dissipating effectors appear to be suppressed at night, a response that contributes to the maintenance of an elevated core temperature (Shido, 1987). Behaviorally, the reinforcement for radiant heat in a cold environment in ovariectomized rats is higher during the day than at night (Carlisle et al., 1979). Briese (1985) first reported that the selected T_a for rats in a temperature gradient was reduced during the night. Furthermore, a recent study has found clear reductions in selected T_a for rat and hamster during the night (Figure 5.7). The selection of cooler temperature at night is well correlated with the increase in motor activity (Gordon, 1993).

The evidence in nocturnal rodents indicates that during the active period at night, autonomic effectors respond to elevate the core temperature, while behavioral effectors exert an opposite action. These responses confound an understanding of the mechanism of the nighttime elevation in core temperature. Pharmacological studies indicate that the CTR involves a regulated elevation in temperature. Administering antipyretics such as indomethacin and salicylates will block the nocturnal rise in core temperature, but has no effect on

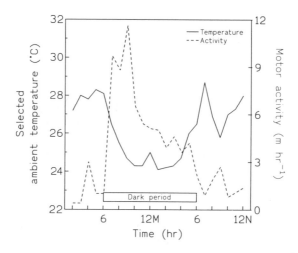

Figure 5.7. Example of the preference for cooler ambient temperatures concomitant with increased motor activity during the dark period of the circadian cycle in the rat. Data from Gordon (1993).

the daytime temperature of the rat (Scales and Kluger, 1987). Because antipyretics block the synthesis of PGE_2, which is thought to be a mediator of fever (see Chapter 2), it has been proposed that the CTR is controlled through an elevation in set-point at night. On the other hand, Briese (1985) has argued that the reduction in selected T_a of the rat refutes the proposed circadian-mediated shift in set-point during the night. That is, an increase in selected T_a would normally be expected if the set-point were elevated (see Chapter 1). Moreover, if the selected T_a approximates the animal's thermoneutral zone during the light phase, then a further increase during the night might lead to excessive heat stress (Gordon, 1993). To clarify these issues, further studies are needed regarding the activation of behavioral and autonomic effectors during crucial periods of the CTR.

5.5.2. Regulation of core temperature during sleep

One aspect of the CTR that deserves special attention is the control of body temperature during sleep. Most of the rodent research in this area has been done with rats. During a typical wake–sleep transition, heat production is maximal during wakefulness, then decreases gradually during synchronized sleep, and is minimal during paradoxical sleep (PS) or rapid-eye-movement sleep (Schmidek, Zachariassen, and Hammel, 1983). The bradymetabolic effect of sleep is most pronounced at cooler ambient temperatures. For ex-

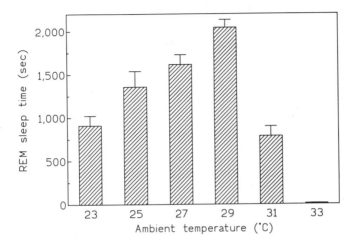

Figure 5.8. Effects of ambient temperature on amount of time spent in rapid-eye-movement (REM) sleep by rats. Note that the amount of REM sleep at 29°C is double that at 23°C. Data from Szymusiak and Satinoff (1981).

ample, negative heat storage occurs consistently during PS at ambient temperatures of 15–30°C, whereas heat storage is positive at 35°C. Brain temperature also decreases during the transition from wakefulness to slow-wave sleep (SWS) (Heller and Glotzbach, 1977), whereas during PS there is an elevation in cerebral blood flow and a transient increase in brain temperature (Roussel, Dittmar, and Chouvet, 1980; Obál et al., 1985).

The sleep–wake patterns of rodents are highly interactive with thermoregulatory processes and have only recently been studied in detail. Several studies have shown marked effects of changes in ambient temperature and skin temperature on the control of PS in rodents (Schmidek et al., 1972; Szymusiak et al., 1980; Szymusiak and Satinoff, 1981). For example, the amount of time spent in PS in the rat peaks at an ambient temperature of 29°C and decreases at temperatures above and below this optimal temperature (Figure 5.8). Interestingly, the amount of time spent in SWS is unaffected over an ambient-temperature range of 23–33°C. Similarly, the maximum PS time peaks at 30°C in the golden hamster, but the optimal ambient temperature for PS is lowered to 20–25°C following several weeks of cold acclimation (Sichieri and Schmidek, 1984). It should be noted that the temperature of peak PS time has been found to lie in the middle of the metabolic thermoneutral zone in many studies of the rat and hamster (see Chapter 3). This could be important to a variety of psychophysiological disciplines, because the proportion of time dedicated to PS has been shown to affect a variety of key psychological parameters (e.g., Szymusiak and Satinoff, 1981).

Although many studies have focused on the impact of thermal stress on the sleep–wake patterns, it has recently been noted that forced changes in sleeping behavior may also affect thermoregulation. For example, sleep deprivation for 24 hr in the rat is associated with an elevation in brain (e.g., cortex) temperature. Furthermore, during recovery from sleep deprivation there is a prolonged reduction in body temperature (Franken et al., 1991).

5.6. Effects of psychological stress on body temperature

Thermoregulation is one of many systems that are exquisitely responsive to stress. Core temperatures of rodents change in the face of a variety of environmental stresses. Obviously, the impact of stress on body temperature is an important topic because the resultant changes in the activities of thermoregulatory effectors and body temperature can affect the results of a variety of experimental procedures.

Animal restraint is a common procedure that often is unavoidable in a variety of research protocols. It is not surprising that thermoregulation is altered in a restrained rodent, especially under conditions of heat or cold stress. The heat balance in restrained rodents is compromised because they cannot display normal thermoregulatory behaviors such as huddling, adjustments in posture, and grooming. Most restraint devices insulate the animal, which further limits heat-exchange processes. Most importantly, restraint can be immensely stressful and can lead to activation of the hypothalamic-pituitary-adrenal axis and associated responses, which in turn affect heat balance.

In a rat restrained in a plastic holder at an ambient temperature of 25°C, core temperature increases by over 1.0°C, peaking after ~30 min of restraint and not reaching the prerestraint core temperature for at least 4 hr (Morley et al., 1990). Interestingly, releasing the rat after 4 hr of restraint is associated with another transient elevation in core temperature. Nagasaka et al. (1979) assessed the effects of physical restraint on total-body heat exchange in rats maintained at 25°C. A naive rat placed in restraint exhibits a marked elevation in heat production, but without an equivalent increase in heat loss, resulting in positive heat storage and a 0.6°C elevation in core temperature (Table 5.8). The restraint-mediated increase in metabolic rate is partly attributable to an activation of BAT thermogenesis. Sectioning the sympathetic nerves innervating BAT practically eliminates restraint-induced BAT thermogenesis and also attenuates the elevation in core temperature (Shibata and Nagasaka, 1982). Restraint-induced hypermetabolism can also be attenuated

Table 5.8. *Effects of restraint on heat production (M), heat loss (H_t), heat storage (S), and colonic temperature (T_{col}) in rats[a]*

Condition	M (W m^{-2})	H_t (W m^{-2}	S (W m^{-2})	T_{col} (°C)
Prerestraint				
5 min[b]	52.4	53.9	−1.5	37.4
40 min	53.9	54.6	−0.7	37.4
Restraint				
5 min	69.1	54.0	15.1	37.7
40 min	68.6	67.4	1.1	38.0
140 min	71.7	68.1	3.6	38.1
Postrestraint				
5 min	57.5	63.3	−5.7	38.1
40 min	54.7	60.9	−6.2	38.0
80 min	57.2	57.1	0.06	37.8

Source: Adopted from Nagasaka et al. (1979).
[a]Note the increase in heat storage during the initial stages of restraint, and the negative heat storage during the postrestraint period.
[b]Time in each condition.

by adrenalectomy or chemical sympathectomy via administration of 6-hydroxydopamine (Nagasaka et al., 1979). This is not surprising considering that plasma levels of epinephrine can increase by 20-fold over basal values in restrained rats (Buhler et al., 1978). Overall, it appears that restraint-induced hyperthermia is derived from a general sympathoadrenal response as well as an activation of BAT thermogenesis.

Whereas restraint causes hyperthermia at ambient temperatures ≥ 25°C, restraint at relatively cool temperatures can result in hypothermia. For example, a naive rat restrained and exposed to 2°C becomes hypothermic, an effect primarily attributed to an inability to shiver (Shimada and Stitt, 1983). However, if the rats are adapted to the restraint procedure, they are eventually capable of shivering and remain normothermic. A cold-exposed guinea pig placed in restraint exhibits a large increase in metabolic rate but still becomes hypothermic, presumably as a result of increased heat loss (Bartlett, 1959). It is clear that as long as restraint is used in rodent studies it will be important to understand the mechanisms of thermoregulation, including overall heat balance, of naive and adapted animals when placed in restraint devices.

There is some recent evidence to suggest that stress-induced changes in body temperature involve changes in the set-point. A variety of psychological stresses, such as placing a rat in an open field or in a cage previously occupied by another rat, will lead to more than a 1.0°C rise in core temperature

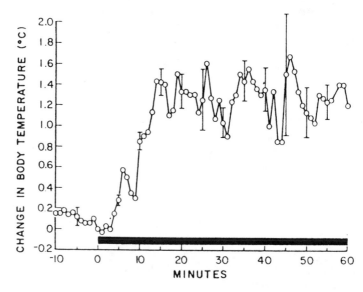

Figure 5.9. Response of core temperature in a rat to psychological stress (black bar) involving placement in an open-field environment (a well-illuminated box with white interior, measuring 4 × 4 × 2 feet). Reprinted from Singer et al. (1986) with permission from Pergamon Press.

(Figure 5.9). This stress-induced hyperthermia can be blocked by preadministration of antipyretics such as indomethacin and sodium salicylate (Kluger et al., 1987; Long, Vander, and Kluger, 1990). Moreover, Briese and Cabanac (1991) have found that stress-induced hyperthermia during the day and night in rats is accompanied by vasoconstriction of the tail, further supporting the notion that the hyperthermia is regulated. A mouse will also exhibit a rise in core temperature following "handling" stress; however, the rise in temperature is not blocked by preadministration of salicylate (Cabanac and Briese, 1991). Taken together, the data suggest that a major component of the hyperthermia during stress in rats is mediated through prostaglandin synthesis in the CNS, suggesting that the hyperthermia is mediated through the same neural pathways as is fever. Stress-induced hyperthermia may operate through different pathways in other rodent species, such as the mouse.

6

Growth, reproduction, development, and aging

Nearly every facet of an organism's life, beginning with the formation of gametes and progressing through conception, prenatal and postnatal development, aging, morbidity, and death, is affected in one of several ways by temperature. The thermoregulatory system must adapt over the continuum of morphological and physiological changes in life in order to efficiently maintain thermal homeostasis. There also are critical periods in development during which the thermoregulatory system is extremely vulnerable to environmental perturbations, thereby creating unstable conditions for other homeostatic processes. The purpose of this chapter is to discuss the effects of environmental temperature and body temperature on crucial developmental aspects of laboratory rodents, including growth, reproductive function, prenatal and postnatal development, and aging.

6.1. Optimal thermal conditions for growth

The thermal limits for optimal growth are key variables in successful management of animal colonies. It is important to know how various aspects of growth and development, such as weight gain, organ development, fertility, fecundity, and fetal development, are affected by subtle changes in environmental temperature. That is, do these variables have a zone of thermoneutrality (see Chapter 3) or a range of ambient temperatures in which their functioning and development are optimal? Furthermore, does the optimal temperature for a given variable change with age? Are the optimal zones of development equivalent between species?

Yamauchi et al. (1981, 1983) performed a thorough study of the effects of graded changes in ambient temperature, ranging from 12°C to 32°C, on a multitude of growth and developmental variables in two generations of rats and mice (Table 6.1). Some variables in the rat have relatively narrow

Table 6.1. *Summary of ambient temperatures above and below which various developmental and growth parameters are significantly affected in the first and second generations (gen) of male (M) and female (F) rats and mice*

	Ambient temperature limits (°C)	
Parameter	Lower	Upper
Rat		
Body weight gain (4–9 weeks) (M)	12	30
Body weight gain (4–9 weeks) (F)	16	28
Delivery rate (1st litter)	12	30
Delivery rate (2nd litter)	12	30
Litter size (1st litter)	12	28
Litter size (2nd litter)	12	30
Weaning rate (1st litter)	12	28
Weaning rate (2nd litter)	14	30
Body weight (3 weeks) (1st litter)	18	28
Body weight (3 weeks) (2nd litter)	16	28
Body weight gain of 1st litter (4–10 weeks) (M)	18	28
Body weight gain of 1st litter (4–10 weeks) (F)	20	26
Food intake of 1st litter at 5 weeks (M)	14	28
Food intake of 1st litter at 5 weeks (F)	16	26
Water intake of 1st litter at 5 weeks (M)	12	26
Water intake of 1st litter at 5 weeks (F)	12	26
Liver weight of 1st litter at 12 weeks (M)	18	32
Liver weight of 1st litter at 12 weeks (F)	18	32
Kidney weight of 1st litter at 12 weeks (M)	18	32
Kidney weight of 1st litter at 12 weeks (F)	18	32
Mouse		
Body weight at 24 weeks (M)	18	28
Body weight at 24 weeks (F)	14	28
Delivery rate (1st litter)	12	32
Delivery rate (2nd litter)	12	28
Litter size (1st litter)	12	26
Litter size (2nd litter)	12	26
Weaning rate (1st litter)	12	30
Weaning rate (2nd litter)	12	26
Body weight at 3 weeks (1st litter; 2nd gen)	14	26
Body weight at 8 weeks (M; 2nd gen)	12	26
Body weight at 8 weeks (F; 2nd gen)	12	26
Food intake at 8 weeks (M; 2nd gen)	18	28
Food intake at 8 weeks (F; 2nd gen)	18	28
Water intake at 8 weeks (M; 2nd gen)	16	26
Water intake at 8 weeks (F; 2nd gen)	18	26
Erythrocytes at 10 weeks (M; 2nd gen)	20	30
Erythrocytes at 10 weeks (F; 3rd gen)	20	28
Leukocytes at 10 weeks (M; 2nd gen)	20	26
Leukocytes at 10 weeks (F; 2nd gen)	20	26
Liver weight at 10 weeks (M; 2nd gen)	20	32
Liver weight at 10 weeks (F; 2nd gen)	18	32
Kidney weight at 10 weeks (M; 2nd gen)	14	28
Kidney weight at 10 weeks (F; 2nd gen)	18	32
Heart weight at 10 weeks (M; 2nd gen)	14	30
Heart weight at 10 weeks (F; 2nd gen)	18	28

Source: Data from Yamauchi et al. (1981, 1983).

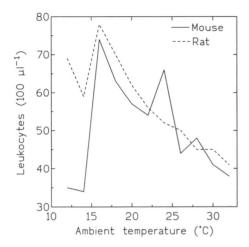

Figure 6.1. Effects of ambient temperature on circulating levels of leukocytes in rats and mice. Data from Yamauchi et al. (1981, 1983).

thermal zones, such as body weight gain for litters of pups (20–26°C), whereas others have wide zones, such as body weight gain for adults, delivery rate of pups, and litter size. Organ weights are remarkably stable over a temperature range of 12–32°C. Most of the growth, developmental, and organ-weight responses of mice are similar to those in rats (Table 6.1). For both mouse and rat the level of circulating leukocytes increase with decreasing ambient temperature, reaching a maximum at 15°C (Figure 6.1). From the results of such studies one can conclude that an ambient-temperature range of 20–26°C is optimal for growth and development in the mouse and rat. It should be noted that the ambient temperature in the typical single-rat plastic cage with wood shavings for bedding is generally 1–2°C higher than its surrounding environmental temperature.

There are considerable discrepancies in the published studies regarding the optimal conditions for growth in the rat; recommended optimal ambient temperatures for growth and development range from 19°C to as high as 29°C. (Yamauchi et al., 1981, 1983). It should be borne in mind that the recommended range of 20–26°C for the rat falls below the lower critical temperature for elevating the metabolic rate (see Chapter 3). By raising the ambient temperature to the upper range of ~26°C, metabolic rate and consequently food consumption are reduced, thus providing economic benefit. However, at these warmer temperatures mice and rats have higher water intake, resulting in greater mineral requirements and increased susceptibility to disease

(Weihe, 1965). Thus, it has been recommended that for mice and rats the ambient temperature be maintained at a relatively cool level of 20–22°C.

It is interesting to correlate a rodent's behavioral thermal preference with its optimal thermal requirements for growth. The rat generally selects ambient temperatures in the same range as that for optimal growth (e.g., 20–26°C). On the other hand, the mouse, which has similar thermal growth requirements, selects markedly warmer temperatures in the range of 30–32°C (see Chapter 4). The degree of matching between a species' thermoregulatory behavior and its thermal zone for optimal growth and development is an interesting paradigm of homeostasis that deserves further study.

6.2. Effects of thermal stress on reproductive function

The capacity to produce viable offspring clearly is a most important aspect of ecological fitness. An individual's ability to acclimate to adverse environmental temperatures does not provide a complete picture of the thermal adaptability of the species. That is, if individuals in a population survive indefinitely in a given set of environmental conditions, but are unable to reproduce, then the species is unfit for such an environment. Thus, it is important to assess how thermoregulatory requirements in an adversely warm or cold environment impact on reproductive function.

Male.   Various functions of the gonads, including hormone secretion and production of viable sperm and ova, are highly sensitive to changes in temperature. The effects of ambient temperature on gonadal function can occur at several levels, including direct effects of body temperature on the gonads, as well as indirect effects via the hypothalamic-pituitary axis. Generally, prolonged exposure to a warm environment is more detrimental to reproductive function than is exposure to a cold environment.

The debilitating effects of heat on spermatogenesis in rodents and other species have been well documented (Blackshaw, 1977). There is marked heterothermy between the testes and abdomen in the rat, and presumably other rodents. This temperature differential is critical to normal reproductive function and is highly sensitive to ambient temperature. For example, at an ambient temperature of 24°C, the temperature of a rat's testis is 3–4°C below its abdominal temperature, whereas the temperature of the caput epididymis is 1°C higher and that of the cauda epididymis is 3–4°C lower than that of the testis (Brooks, 1973). Thus, spermatozoa leaving the testes are subjected to an increase in temperature in the caput and then a decrease in the cauda epididymis. The testicular–abdominal temperature gradient of the rat is quite la-

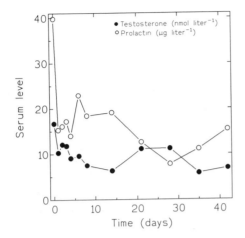

Figure 6.2. Effects of continuous exposure to 33–35°C on serum levels of prolactin and testosterone in the male rat. Data from Chap and Bedrak (1983).

bile and can be quickly attenuated to <1°C with a brief exposure to an ambient temperature of 37°C, but increased to >8°C by exposure to 5°C. The temperature gradient can also be modified by adjustments in posture (Brooks, 1973). Because postural adjustments are commonly used thermoregulatory behaviors in rodents (see Chapter 4), it would seem that the effects of ambient heat and cold stress on the male reproductive tract could be influenced by both behavioral and autonomic thermoregulatory processes.

Prolonged exposure to temperatures above thermoneutrality is generally detrimental to reproductive function in the male rodent. For example, when the male rat is acclimated to 35°C, the resultant elevations in scrotal and core body temperatures (1.3°C and 0.9°C, respectively) are sufficient to cause marked reductions in frequency of mating and percentage of conceptions, as well as degeneration in the seminiferous tubules (Sod-Moriah, 1974). The elevation in scrotal temperature during heat acclimation is not the only factor contributing to decreased fertility in the male rat. Prolonged heat stress also affects the hypothalamic-pituitary-testicular axis, which further contributes to a compromise in reproductive function. Continuous exposure to 33–35°C in the rat leads to reductions in serum concentrations of prolactin and testosterone (Figure 6.2). These serum hormonal changes occur simultaneously with reductions in follicle-stimulating hormone (FSH), luteotropic hormone (LH), and prolactin in the adenohypophysis. It appears that the reduced serum prolactin is a major factor leading to a reduction in the synthesis of testosterone (Chap and Bedrak, 1983).

The pineal gland, in addition to its well-known function as a neuroendocrine transducer of photic stimuli, serves a crucial role in the modulation of reproductive function in rodents subjected to chronic heat stress. It appears that the pineal may protect the testes of the hamster (and perhaps the rat) from damage during prolonged exposure to high temperatures. In the rat pineal, the activity of N-acetyltransferase, the enzyme involved in melatonin synthesis, is markedly attenuated after just 3 days of exposure to 33°C (Nir and Hirschmann, 1978). Chronic heat stress lowers melatonin production by the pineal gland, whereas complete removal of the pineal gland exacerbates the debilitating effects of heat. For example, the pinealectomized hamster maintained at 34°C exhibits reduced testicular weight and damaged seminiferous tubules, along with reduced pituitary LH and prolactin levels (Kaplanski et al., 1983). Similarly, the pinealectomized, heat-exposed rat shows reductions in pituitary LH and serum testosterone levels; however, the deficits appear not to be as deleterious as those observed in the hamster (Magal et al., 1981). The pineal gland's protective effect on testicular function during heat stress is an enigma. Melatonin production is reduced during heat stress, an effect that should lessen the debilitating effect on the testes. Yet total removal of the pineal gland appears to worsen the effects of heat stress.

Female.   The reproductive functions of female rodents appear to be as strongly affected as those of the males during prolonged exposure to heat stress. The estrous-cycle duration is increased in rats following acclimation to 35°C (Sod-Moriah, 1971). Heat acclimation also results in higher serum progesterone levels during diestrus and during the first 5 days of pregnancy in rats (Bedrak et al., 1977). The number of corpora lutea is unaffected by heat acclimation, but the percentages of fertilized and implanted ova, the number of metrial glands, and the number of young born as a percentage of the corpora lutea are all reduced in heat-acclimated rats.

Rats have smaller litters and earlier weaning times when reared at ambient temperatures of 30–32°C, which, it should be noted, is at the upper end of the thermoneutral zone (Yamauchi et al., 1981). Rats that survive an extreme core-temperature elevation of 42.5°C still are able to reproduce normally at a later time, although there is a possibility of increased sterility in male rats subjected to these temperatures (Furuyama, Ohara, and Ota, 1984). Overall, it seems that rats begin to show adverse signs regarding reproductive function when maintained at ambient temperatures ≥30°C.

Hamsters maintained at 34°C have normal ovarian weights, but reduced numbers of corpora lutea, litter sizes, and litter weights (Kaplanski et al.,

1988). Pinealectomized hamsters show greater susceptibility to heat exposure, as indicated by reduced ovarian weight. Just as it does in rats, pinealectomy accentuates the deleterious effects of chronic heat stress on various reproductive parameters in the hamster (Magal et al., 1983). Thus, production of melatonin by the pineal gland provides protection to the reproductive function in male and female rodents. Paradoxically, keeping the female hamster at 30°C instead of 22°C affords protection against melatonin-induced suppression of reproductive function (Reiter et al., 1988).

The effects of cold exposure on reproductive function have not been as widely studied as those of heat exposure. Relatively mild cold exposure (e.g., $T_a \geq 12°C$) with unlimited availability of food appears to pose marginal adversity to reproductive function in rodents (Yamauchi et al., 1981). On the other hand, the energy requirements of thermoregulation in extremely cold environments may impact the functioning of reproduction and other physiological processes (Barnett, 1973; Manning and Bronson, 1990). The estrous cycle of the golden hamster begins to deteriorate after 15 days of exposure to 7.5°C, and complete anestrus is achieved by 51 days of cold exposure (Grindeland and Folk, 1962). Mice given food ad libitum will exhibit normal ovulatory cycling and uterine growth at ambient temperatures as low as −8°C; however, restricting their food intake, and thus forcing the mice to rely more on their internal fat stores for energy balance, results in uterine atrophy and cessation of ovulation at 12°C (Figure 6.3). This response provides an interesting paradigm of the interaction between the primary energy demand for thermoregulation and a secondary demand for functioning of the ovaries and accessory reproductive organs; it appears that when the former is limited, the latter is sacrificed to ensure survival in the cold environment.

6.2.1. Thermoregulation during pregnancy and lactation

The maternal energy requirements for development of the fetuses and newborns also place marked restrictions on thermoregulation. In the preceding section it was mentioned that gonadal function is often compromised in extremely cold environments in order to assure an adequate energy store for thermoregulation. However, after conception, one generally sees an attempt to compromise certain thermoregulatory responses in order to meet the nutritional and metabolic needs of the fetuses. In this regard, the activity of the sympathetic-nervous-system–BAT axis becomes particularly affected during pregnancy.

The nursing of offspring requires a tremendous amount of energy expenditure. A reduction in BAT thermogenesis is a key adaptive response to

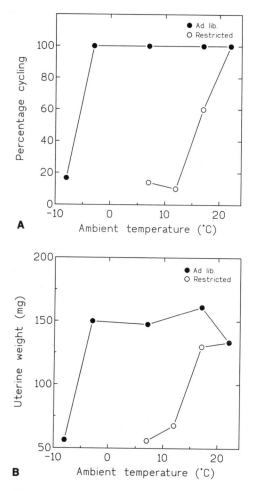

Figure 6.3. Interactive effects of ambient temperature and food availability on reproductive function in female mice as measured by the percentage of animals with normal ovulatory cycling (A) and normal uterine weight (B). Mice with limited caloric
intake in the cold cannot maintain a normal reproductive function. Data from Manning and Bronson (1990).

economize on maternal energy expenditure during lactation (Trayhurn,
1985). Indices of BAT function, such as BAT weight and GDP binding, decline markedly during pregnancy and lactation in the mouse, golden hamster,
and rat (Isler, Trayhurn, and Lunn, 1984; Andrews et al., 1986; Wade, Jennings, and Trayhurn, 1986; Trayhurn and Wusteman, 1987). By day 15 of
pregnancy in the hamster, the cytochrome oxidase activity in BAT is reduced

Table 6.2. *Influence of stage of pregnancy on thermal and metabolic responses to cold exposure* ($T_a = 10°C$ *for 30 min) in rats*

Stage[a]	T_c at 25°C	M at 25°C (ml min^{-1} kg$^{-0.67}$)	$-\Delta T_c$ (°C at 10°C)	ΔM at 10°C (ml min^{-1} kg$^{-0.67}$)
Virgin	38.5 ± 0.1	23.2 ± 1.3	0.4 ± 0.2	15.2 ± 0.9
Pre 1 week	37.7 ± 0.1	22.1 ± 1.6	0.8 ± 0.3	9.6 ± 1.1
Pre 0	37.5 ± 0.2	20.2 ± 1.8	1.8 ± 0.3	4.5 ± 1.0
Post 0	39.0 ± 0.2	18.5 ± 1.3	0.3 ± 0.3	13.8 ± 1.8
Post 1 week	39.0 ± 0.1	25.4 ± 2.1	0.4 ± 0.2	11.1 ± 2.0

Source: Adapted from Imai-Matsumara et al. (1990).
[a]Pre 1 week, 1 week prior to parturition: pre 0, 24 hr prior to parturition; post 0, 24 hr after parturition; post 1 week, 1 week after parturition.

by approximately 50% compared with that in virgin animals (Quek and Trayhurn, 1990). BAT sensitivity to norepinephrine (NE) is suppressed during pregnancy and lactation in the rat (Villarroya, Felipe, and Mampel, 1987) and during lactation in the mouse (Trayhurn, 1983), but returns to normal following weaning. It is interesting to note that despite the hyperphagia associated with pregnancy, diet-induced thermogenesis is also suppressed (Abelenda and Puerta, 1987).

There is a novel thermal effect of ambient temperature on cannibalistic behavior by the lactating dam. Syrian hamsters will cannibalize some of their young during periods of energy deprivation. When hamsters are maintained at an ambient temperature of 10°C, they will eat significantly more offspring during postnatal days 1–8, a response that apparently evolved as a means to conserve the dam's energy stores during lactation (Schneider and Wade, 1991). Thus, in this unusual scenario, some of the young are sacrificed in order to meet the additional energy demands for thermoregulation in the dam and to ensure survival of at least some of the nursing offspring.

Changes in nonshivering thermogenic function during pregnancy and lactation impact on various thermoregulatory processes, which show abrupt changes in function prior to and after parturition. For example, the thermogenic and body-temperature responses of pregnant rats to cold exposure are markedly affected by the stage of pregnancy (Table 6.2). An impaired thermogenic response to cold exposure is seen 24 hr prior to parturition, followed by a remarkable recovery by 24 hr after parturition. Moreover, during the 24-hr period prior to parturition, rats are especially susceptible to hypothermia when cold-exposed. Such a depression in metabolic sensitivity could be beneficial to the fetus, because a normal, sympathetic-mediated elevation

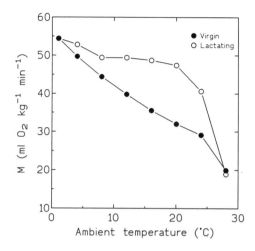

Figure 6.4. Example of change in thermoregulatory profile during lactation in the rat. The lactating animal had a markedly higher metabolic rate over a temperature range of 12–24°C. Data from Roberts and Coward (1985).

in thermogenesis during cold exposure can cause vasoconstriction in the placenta and compromise fetal blood flow (Imai-Matsumara et al., 1990). Thus, at the risk of becoming hypothermic, late-term pregnant rats continue to maintain a high rate of placental blood flow during cold exposure.

The temperature–metabolism profile below the lower critical temperature for the rat is also altered during lactation (Figure 6.4). At thermoneutrality (28°C), metabolic rates are similar in virgin animals and lactating animals; however, as temperature is lowered, the metabolic rate of the lactating animal is maintained at a much higher level. Considering that BAT thermogenesis is most likely suppressed, it is not clear what causes the metabolic rate of the lactating rat to increase at cooler temperatures. The increase in metabolism likely is attributable to sources of nonshivering thermogenesis other than BAT and/or shivering. In this regard, it is also of interest to note that lactating rats have higher core temperatures at ambient temperatures of 4–28°C and are more susceptible to hyperthermia during contact bouts with their rat pups (Jans and Leon, 1983; Adels and Leon, 1986). Clearly, we need a better understanding of the behavioral and autonomic thermoregulatory responses of lactating rodents exposed to warm and cold environments.

Other thermoregulatory parameters are also altered during pregnancy and lactation. For example, the amplitude of the circadian temperature rhythm decreases during pregnancy and then increases during lactation in rats (Kittrell and Satinoff, 1988). The threshold core temperature for salivary secretion and the ambient temperature for grooming saliva are significantly reduced in pregnant rats (Wilson and Stricker, 1979). A pregnant rat exposed to heat stress will attempt to maintain its elevated defended temperature (see Chapter 5) at a lower level than will a virgin rat. This may represent an adaptive response to protect the developing fetuses from hazardous temperatures. Overall, heat tolerance is markedly reduced during pregnancy, perhaps attributable to an impaired ability to maintain sufficient rates of evaporative cooling (Knecht, Toraason, and Wright, 1980).

6.2.2. Effects of thermal stress on fetal development

Fetal development is exquisitely sensitive to environmental perturbations. Homeostasis of temperature and other variables during pregnancy is crucial for normal fetal development. A breakdown of thermal homeostasis in the dam, particularly in hot environments, can result in a number of pathological sequelae in the fetus and neonate.

The work of M. J. Edwards has been pivotal in determining the effects of maternal hyperthermia on the incidence of fetal teratogenesis in rodents (Edwards, 1969, 1982; Germain, Webster, and Edwards, 1985). Hyperthermia-induced teratogenesis is dependent on several variables, particularly the duration and magnitude of the maternal elevation in temperature and the period of gestation during which heat exposure takes place. For example, at 9.5 days of gestation, a stage at which the rat embryo is most sensitive to heat, an elevation in core temperature of 2.5°C for 60 min induces teratogenic sequelae (Germain et al., 1985). The threshold embryonic temperature that retards brain development in rats is only 1.5–1.7°C above baseline (Edwards, 1982), but the threshold maternal core temperature for inducing fetal death and/or malformations is 41.5°C, when heated by exposure to 27.12-MHz radio-frequency radiation (Lary et al., 1986).

Other rodent species appear to be equally sensitive to heat stress. As in the rat, the sensitivity of the pregnant hamster to heat varies with the period of gestation. Fetal resorptions predominate at day 8, whereas fetal malformations occur mostly on day 9 of gestation (Umpierre and Dukelow, 1978). Fetal malformations will occur in the hamster at day 8 of gestation on exposure to an ambient temperature of 40°C for as little as 40 min (Kilham and Ferm, 1976; Umpierre and Dukelow, 1978). Among mice, significant defects and embryo mortality can be detected when pregnant females are exposed during the time of the first cleavage division to an ambient temperature of 34°C for just 24 hr (Elliott, Burfening, and Ulberg, 1968; Bellve, 1973). These embryonic defects occur with maternal temperature elevations of only 2.0°C.

Combinations of exposure to temperatures of 30–36°C and exercise will result in slight but significant reductions in fetal weight of mice and are associated with core temperatures of only 38–40°C (Berman, House, and Carter, 1990). Edwards (1982) estimated that there was an 8.4% decrease in the brain weight of the newborn guinea pig for each 1°C elevation in maternal core temperature above 41°C (1-hr exposure on day 21 of gestation). Overall, notwithstanding the subtle fetal defects, it can be stated that the threshold maternal core temperature for eliciting acute teratogenic symptoms is approximately 41°C. Significant changes in fetal development can also be observed with more subtle elevations in core temperature.

It is also evident that acute cold stress can elicit adverse changes in fetal development. For example, exposure of pregnant rats to 4°C for nearly the entire duration of gestation results in lower body weights for the newborns (Saetta, Noworaj, and Mortola, 1988). Lowering the core temperature of pregnant mice to 20°C for 10 hr causes fetal resorption and abortion, dead and abnormal fetuses, and failure of implantation (Stankiewicz, 1974). These effects are most pronounced on days 6–13 of gestation, whereas there is little effect from this severe state of hypothermia on day 20 of gestation. It is likely that the teratogenic effects of cold stress occur as a result of maternal stress, rather than having a direct thermal effect on the fetus as occurs with hyperthermia.

6.3. Development of thermoregulation from birth to weaning

Study of the development of thermoregulation in rodents is one of the most dynamic areas in the physiological sciences. Indeed, within a period of 3–4 weeks, altricial, poikilothermic rodent neonates will develop the ability to thermoregulate over a relatively broad range of ambient temperatures. During this relatively brief period between birth and weaning there is a wide array of behavioral and autonomic changes in the neonates and dam that play major roles in the development of homeothermy; for review, see Satinoff (1991).

There is considerable interspecies variation in the stability of the thermoregulatory systems of neonatal rodents. This is attributable in part to the species' gestation periods: only 16 days for hamster, 20 days for mouse, 22 days for rat, 24 days for gerbil, and 67 days for guinea pig. The species with short gestation periods (mouse, gerbil, hamster, and rat) are virtually poikilothermic at birth and become fully homeothermic over a period of approximately 20–30 days. The newborn guinea pig is precocial and has relatively well developed thermoregulatory control, which is a reflection of its long gestation period. In the smaller rodent species there are marked changes

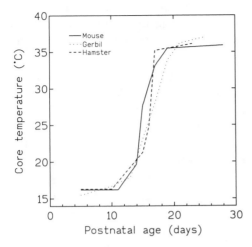

Figure 6.5. Stabilized core temperatures of preweanling mouse, gerbil, and hamster exposed to an ambient temperature of 14–16°C. Note marked development of thermal homeostasis at an age of 15 days. Data from Lagerspetz (1966)(mouse), McManus (1971)(gerbil), and Hissa and Lagerspetz (1964)(hamster).

in various behavioral and autonomic motor outputs that can be detected as early as postnatal day 1, as discussed later; however, the ability to maintain thermal homeostasis over a relatively wide range of environmental temperatures in these smaller rodent species is not evident until the end of the second week of development and is relatively well developed by the time of weaning (Figure 6.5). After weaning, the ability to thermoregulate in hot and cold environments gradually improves as the animal's body size increases, leading to a reduction in whole-body thermal conductance (see Chapter 3). Overall, the development of cold resistance in the rat as a function of age can be grouped into five general periods: (1) birth to 18 days, when there is relatively little cold resistance as compared with the adult; (2) 18–30 days, a period of rapid development of cold resistance; (3) 31–60 days, a period with marginal improvements in cold resistance; (4) 61–300 days, the period of maximal cold resistance; (5) 300 days to death, a period associated with slow deterioration in cold resistance (Hill, 1947). For a most thorough analysis of the development of thermoregulation and other physiological systems in the rat, the reader is referred to the landmark review by E. F. Adolph (1957).

Behavioral thermoregulatory responses. Measuring behavioral and autonomic thermoregulatory responses in newborn rodents presents challenges not encountered with adults. Because of their small size and lack of insulation, newborn rodents easily cool when kept individually at temperatures below their lower critical temperatures, which are well above 30°C for the mouse, gerbil, hamster, and rat. Thus, to avoid severe hypothermia and its resulting depressant effects on behavior and physiology in neonates, one must consider ambient temperature, the presence of insulative bedding in the test chamber, and the number of animals in a test group (i.e., huddling) when measuring physiological and behavioral responses of newborn rodents.

Huddling with their littermates and with the dam is crucial for thermal homeostasis in newborns of the altricial rodent species. Alberts (1978) demonstrated that the huddling behavior of rat pups is complex, being affected by environmental temperature, the number of pups in a group, and other factors. For example, at a relatively cool ambient temperature of 20°C, a litter of 5-day-old pups will huddle in a clump to reduce the total exposed surface area of the group. As temperature is increased to 38–40°C, the litter will not huddle as closely, thereby increasing the total exposed surface area of the group to facilitate heat loss.

Other forms of thermoregulatory behavior, such as thermotaxis, can also be detected in neonates. A 1-day-old rat pup will display thermotaxis in a temperature gradient provided that the initial temperature at placement in the gradient is not so cold that the pup cools too quickly and is rendered hypothermic and unable to move in the gradient (Kleitman and Satinoff, 1982). As the rat pup ages, it selects warmer temperatures in a gradient, resulting in a higher core temperature (Fowler and Kellog, 1975). Studying thermoregulatory behavior in newborn animals can also be confounded by psychological factors not encountered in adults. For instance, the thermotactic behavior of a very young rat pup can be influenced by deprivation and handling, effects that should be considered when interpreting thermoregulatory behavior in neonatal rodents (Johanson, 1979).

In a temperature gradient the selected T_a of the newborn gerbil reaches a maximum of 37.9°C at 5 days of age and then decreases gradually, reaching a nadir of 30.1°C at 12 weeks of age (Eedy and Ogilvie, 1970). The mouse similarly shows a very high selected T_a at an early age that steadily decreases, reaching a nadir of 32.4°C at 4 weeks of age. Newborn hamsters are very immature, with the poorest development of BAT, but they compensate with relatively well developed behavioral thermotactic responses (Leonard, 1974, 1982; Schoenfeld and Leonard, 1985). When the hamster dam leaves the nest, the littermates huddle in the bottom of the nest, reducing the exposed surface area and conserving heat. The newborn hamster displays exquisite thermal orientation behavior when placed in a temperature gradient. This thermotactic behavior declines abruptly at 8 days of age, a response that appears to be related to the pup's development of olfaction (Schoenfeld and Leonard, 1985).

Autonomic responses.   Behavioral responses can make a limited contribution to thermal homeostasis during periods of cold stress, such as when the dam leaves the nest. Otherwise the neonates must rely on autonomic effectors to thermoregulate. With the exception of the guinea pig, newborn rodents have poorly developed musculoskeletal systems and thus are unable to generate heat by shivering. However, they nonetheless have the capacity to respond thermogenically to environmental and pharmacological stimuli. For example, a thermogenic response to environmental cooling can be detected in a rat pup as soon as 4 hr after birth (Várnai and Donhoffer, 1970). A 4-day-old mouse pup will show a clear thermogenic response to ambient cooling and parenteral administration of NE (Figure 6.6).

Because of their inability to shiver, it follows that animals at this early age possess relatively well developed nonshivering thermogenic mechanisms. BAT is highly developed in the newborn mouse, rat, and guinea pig; for reviews, see Skala (1984) and Himms-Hagen (1990a,b). Indeed, the amount of BAT in the fetal mouse and rat increases dramatically beginning around day 17 of gestation with the concentration of uncoupling protein (see Chapter 4), showing a sharp rise at day 18 of gestation (Houštěk et al., 1988). BAT mitochondria acquire the thermogenic potential typical of an adult approximately 2 days before birth. Between postnatal days 0 and 5, mouse BAT cells exhibit profound morphological changes, including a tripling in size, marked increases in the number and size of BAT mitochondria, and enlargement of BAT cell nuclei (Vinter, Hull, and Elphick, 1982). The sensitivity of the rat to the calorigenic effects of NE is greatest at day 1; however, the magnitude of the thermogenic response does not peak until days 4 and 5 (Hsieh, Emery, and Carlson, 1971).

BAT function changes dramatically in the first postnatal week and is quite sensitive to the prevailing ambient temperature. The weight of BAT drops from day 0 to day 5, which reflects the utilization of BAT lipids for thermogenesis during this critical period of development (Tarkkonen and Julku, 1968). Rats born and reared at an ambient temperature of 28°C show a greater decrease in BAT weight than do those raised at 16°C (Mouroux, Bertin, and Partet, 1990). Moreover, the biochemical developments in BAT functioning, such as cytochrome oxidase activity and GDP binding, are markedly

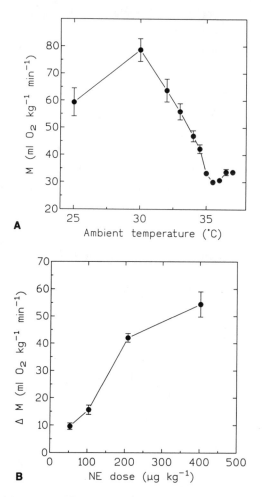

Figure 6.6. Thermogenic responses of individual 4-day-old mouse pups to ambient cooling (A) and subcutaneous administration of NE (B). Note the drop in metabolic rate at 25°C, which reflects the beginning of thermoregulatory failure. Data from Moore and Donne (1984).

increased in rats raised in the cooler environment (Figure 6.7). The newborn hamster, on the other hand, has poorly developed BAT that does not exhibit appreciable function until 12 days of age (Sundin, Herron, and Cannon, 1981).

Generally speaking, the ability of altricial species to shiver develops concomitantly with the depletion of BAT lipid stores. The ability to produce heat by shivering develops in hamsters and rats after approximately 10 days and

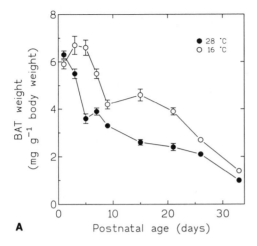

A

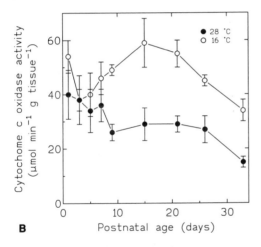

B

Figure 6.7. Time course of BAT weight and BAT cytochrome c oxidase activity in rats born and reared at ambient temperatures of 16°C and 28°C. Note the marked increase in BAT enzyme activity at the age of 10 days. Data from Mouroux et al. (1990).

contributes to the neonate's metabolic heat production. Shivering can be detected in the mouse at an age as early as 5 days, but only when its core temperature decreases below 27°C (Arjamaa and Lagerspetz, 1979). As a mouse pup ages, its threshold core temperature for shivering increases linearly, until it reaches its maximum of 37°C at 20 days of age.

Thermal homeostasis in the newborn guinea pig is relatively well developed compared with the other rodent species discussed here (Farkas, 1978). As in other rodents, the activation of nonshivering thermogenesis in the newborn guinea pig is crucial for its thermal homeostatic capability (Adamsons, Blumberg, and Joelsson, 1969; Holloway et al., 1984). The weight of BAT in the fetal guinea pig increases abruptly at around day 45 of gestation, reaching its peak weight at day 66 (i.e., just prior to birth), and then decreases quickly over the first 15 postnatal days, which reflects reliance on BAT lipids as an energy source (Cresteil, 1977). One interesting demonstration of the advanced development of nonshivering thermogenesis in the guinea pig comes from an experiment by Zeisberger (Zeisberger and Roth, 1988). A guinea pig born prematurely, with closed eyes, sparse hair, and a weight of only 34 g ($\sim$ 33% of normal), was nonetheless capable of nonshivering thermogenesis in response to ambient cooling and NE administration. As the newborn guinea pig ages, shivering plays a greater role in supplying the heat needed for temperature regulation (Brück and Wünnenberg, 1966).

There is a considerable data base on the development of behavioral and autonomic thermoregulatory responses in newborn rats (Table 6.3). Key thermoregulatory variables, such as the thermoneutral zone, the range of normothermy, the maximum metabolic rate, behavioral thermoregulatory responses, and tissue insulation, quickly mature in rats. During the first 3 weeks of development there is a 7°C reduction in the lower critical temperature that is largely attributable to the increase in body size and the development of insulation. The resting metabolic rate at thermoneutrality for individual pups has been found to increase slightly from birth to weaning (Thompson and Moore, 1968; Conklin and Heggeness, 1971; Spiers and Adair, 1986), whereas the maximal oxygen consumption induced by ambient cooling shows a marked increase from 50 to 85 ml O_2 min^{-1} kg^{-1} between postnatal days 6 and 20. The hypothalamothyroid axis of the rat does not mature until approximately 7 weeks of age (Frankel and Lange, 1980).

The overall development of thermoregulatory stability in rodents is incredibly sensitive to changes in temperature. For example, rearing a 2-week-old pup at an ambient temperature of 34°C for only 4 days will reduce its cold resistance by 66% as compared with its littermates reared at normal room temperature (Hahn, 1956). Rearing newborn guinea pigs for 7 days at 36°C has no effect on metabolic rate at thermoneutrality, as compared with animals reared at 21°C, but it markedly lowers the thermogenic response to ambient cooling (Adamsons et al., 1969). The sensitivity of the developing thermoregulatory system to endogenous and exogenous stimuli is a crucial area deserving further study.

Table 6.3. *Summary of the time courses of development for key thermoregulatory variables in the newborn rat*

Variable, dimension Age (days)	Value
Metabolic rate (ml min^{-1} kg^{-1}) (minimum/maximum)[a]	
1	25.2/52.1
3	24.1/53.8
6	24.1/51.5
12	25.2/66.6
20	31.1/83.0
30	33.4/85.3
Lower critical T_a (°C)[b]	
1	35.0
6	34.5
10	34.0
12	33.0
14	32.2
18	32.0
30	28.0
Insulation (°C cal^{-1} min^{-1} cm^{-2})[c]	
0.5	25.0
3	28.0
6.5	49.0
9.5	49.0
14.5	68.0
19.5	102.0
Selected core/skin temperatures when placed in temperature gradient (°C)[d]	
4.5	30.5/28.5
6.5	32.5/30.5
8.5	34.5/32.0
10.5	35.5/33.5
12.5	36.0/34.0

[a]Data from Thompson and Moore (1968).
[b]Data from Thompson and Moore (1968) and Spiers and Adair (1986).
[c]Data from Takano et al. (1979).
[d]Data from Folwer and Kellog (1975).

Circadian temperature rhythm.   The circadian temperature rhythm (CTR) in rats appears to develop during the first week after birth (Schmidt, Kaul, and Heldmaier, 1987). Rat pups isolated from their dams in continuous light continue to display a CTR that begins to fade in the third postnatal week (Nuesslein and Schmidt, 1990). Thus, the juvenile rat exhibits an endogenous CTR that can operate independently of maternal influence. Another study found that the rat CTR did not appear until around day 24 and displayed a maximal response through day 50 (Kittrell and Satinoff, 1988). More work on the development of the CTR in other rodent species is clearly needed, especially comparisons between the guinea pig (a diurnal species) and other rodents.

6.3.1. Hyperthermia-induced seizures in the neonate

It is estimated that as many as 5% of human infants develop a type of convulsion when subjected to excessive body warming during fever. These febrile convulsions are thought to be responsible for some mental deficits that manifest later in life and are considered major problems in pediatric medicine (Holtzman, Obana, and Olson, 1981; MacKintosh, Baird-Lambert, and Buchanan, 1984; Hjeresen and Diaz, 1988). Hence, the development of experimental animal models to study febrile convulsions is a crucial area in developmental physiology.

The term "febrile convulsion," as it applies to studies in which a newborn rodent is heated externally, is somewhat of a misnomer. That is, the rise in core temperature during a fever is distinguishable in several ways from a rise in temperature due to external heat stress (see Chapter 2) – the former being a regulated elevation and the latter a forced elevation in body temperature. Thus, it is more accurate to refer to these convulsions as hyperthermia-induced, rather than febrile.

A preweaning rat displays a pattern of febrile convulsions during whole-body heating that parallels some of the symptoms observed in humans. For example, raising the core temperature of a 2-day-old rat to 37°C will cause a normally quiet animal to roll on its back and experience clonic movements in all four limbs. The threshold temperature for seizures rises dramatically with increasing age of the rat pup, reaching a maximum of 44–45°C at an age of 10 days (Figure 6.8). The hyperthermia-induced seizures are completely reversible up to day 7. After day 10, pups induced into seizures do not recover (Holtzman et al., 1981), although other studies have reported recovery in 20-day-old rat pups after hyperthermia-induced convulsions (McCaughran and Schechter, 1982). It would seem that the extremely low threshold temperature for hyperthermia-induced seizures in the 2-day-old rat pup could make the maternal–newborn contact bouts crucial, especially in the face of elevated temperature. Leon and others (Leon, 1985; Adels and Leon, 1986) have detailed the complex effects of temperature on maternal-newborn contact bouts. How such behavior might affect the prevention of hyperthermia-induced convulsions in very young rat pups (i.e., $\geq$ 2 days) remains to be determined.

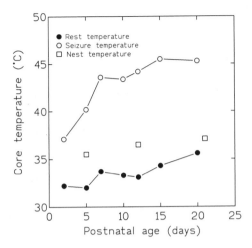

Figure 6.8. Resting core temperature and threshold temperature for eliciting convulsions in individual rat pups. Data from Holtzman et al. (1981). Also plotted is normal core temperature as a function of the ages of groups of rat pups while in the nest. Data from Conklin and Heggeness (1971). Note the rapid increase in the threshold temperature for eliciting seizures from postnatal day 2 to day 7.

Hyperthermic convulsions are elicited in adult mice and rats at core temperatures that are nearly equal to the lethal body temperature (Maxson, 1980; Kasting et al., 1981). It is interesting to note that the homozygous Brattleboro rat, which lacks arginine vasopressin (AVP), and rats immunized to AVP exhibit convulsions at markedly higher core temperatures than do normal rats (Kasting et al., 1981). AVP release appears to be a crucial factor in the causation of hyperthermia-induced and/or febrile convulsions.

6.4. Aging and thermoregulation

Aging involves a gradual regression in the functioning of all physiological systems, including temperature regulation. Indeed, the increased incidences of disease and mortality among elderly humans during seasons of extremely high and low temperatures exemplify the importance of understanding thermoregulatory function in the aged. Aged rodents are used in a variety of fields because a dysfunction or failure of a particular system can be easily detected in an aged animal. Impaired thermoregulatory reflexes can affect the responses of other systems, thus placing greater importance on an understanding of how aging affects thermoregulation. For example, the Morris water swim task is an important tool for studying cognitive functions in rodents;

performance in this task has been found to degenerate in aged rats (Linder and Gribkoff, 1991). However, the Morris task typically is performed at a relatively cool water temperature of ~21°C, which represents cold stress and leads to marked hypothermia in an aged rat. It was found that the impaired cognitive performances of aged rats could be greatly improved when the water temperature was increased to the point that hypothermia was prevented (Linder and Gribkoff, 1991). Indeed, the thermoregulatory system can affect many physiological and behavioral systems that one normally would not expect to be influenced by temperature.

Laboratory rodents, particularly rats, have become valuable experimental models in which to study the progressive failure of thermoregulatory control in the aged. The life spans of laboratory rodents are relative short, ranging from 2 to 3 years. Hence, the effects of age on thermoregulatory functions can be studied over a relatively brief time frame. Most commonly studied are the effects of aging on the central neural control of body temperature, defects in basal thermoregulatory processes, and the ability to increase facultative thermogenesis and thermoregulate during cold exposure.

Central neural effects.   The rat displays various deficits in behavioral and autonomic thermoregulatory control as it approaches an age of approximately 24 months (Table 6.4). Impaired thermoregulation appears to be a result of dysfunctions in motor outputs, as well as their control at the level of the CNS. A 15–18-month-old rat will show greatly reduced sensitivity to intraventricular injection of key neuromodulating agents such as dopamine, PGE_2, and carbachol (Ferguson et al., 1985). Although operant thermoregulatory behavior has been found to be unaffected in aged rats (Jakubczak, 1966), Owen, Spencer, and Duckles (1991) found that cold-acclimated aged rats selected cooler ambient temperatures in a temperature gradient concomitant with lower core temperatures. These few studies suggest that aging-induced deficits in thermoregulation occur at the level of the CNS. Some of the CNS effects could be attributable to dysfunction of the thermoafferent system (see Chapter 2); however, little is known about this area.

Basal thermoregulation.   Basal or resting metabolic rate declines with aging; however, this is attributable in large part to the increased body mass of older animals. Core temperature measured at thermoneutrality has been found to decrease by 0.8°C from the age of 3 to 24 months (Balmagiya and Rozovski, 1983). Mice maintained at 22°C begin to show a progressive decrease in the regulated core temperature at approximately 24 months of age

Table 6.4. *Summary of deleterious effects in the thermoregulatory system in aging rats*

Age (months)	Effect	Reference
22	Impaired ability to adjust behaviorally and autonomically to prolonged exposure to T_a of 6–10°C	Owen et al. (1991)
18	Reduced amplitude of circadian thermoregulatory rhythm	Refinetti et al. (1990)
20–24	Reduced metabolic response to tryamine	Kiang-Ulrich and Horvath (1984b)
21	Reduced survival rate during acclimation to T_a of 5°C	Kiang-Ulrich and Horvath (1985)
23–27	Reduced core temperature and metabolic rate during cold exposure; reduced BAT mitochondrial proteins and GDP binding in males, but not females	Lee and Wang (1985), McDonald et al. (1988a,b, 1989)
22	Altered cardiovascular responses to cold-water immersion	McCarty (1985)
24	Reduced metabolic rate and core temperature at thermoneutrality; reduced metabolic response to NE	Balmagiya and Rozovski (1983)
16–36	Increased fluctuations of tail skin temperatures in females	Simpkins (1984)
24	Reduced ability to thermoregulate during adaptation to restraint and cold stress	Paré (1989)
24–28	Reduced febrile response to *Salmonella typhimurium*	Tocco-Bradley et al. (1985)
15–18	Reduced thermoregulatory responses to intraventricular injection of PGE_2, NE, dopamine, and carbachol	Ferguson et al. (1985)
24	Inability to thermoregulate during heat and cold stress while restrained	Cox et al. (1981)
33–40	Reduced ability to thermoregulate during exposure to T_a of 5°C	Martin et al. (1985)

(Talan, 1984; Reynolds, Ingram, and Talan, 1985). Moreover, there is an inverse correlation between time to death and core temperature in mice, suggesting a close relationship between morbidity and thermoregulatory failure. On the other hand, an increase in core temperature has been reported in the 20- to 24-month-old rat maintained at 23°C (Kiang-Ulrich and Horvath, 1984a). Thermoregulatory stability in the aged mouse and rat is also impaired as temperature is increased above thermoneutrality (Cox, Lee, and Parkes, 1981; Talan and Ingram, 1986a).

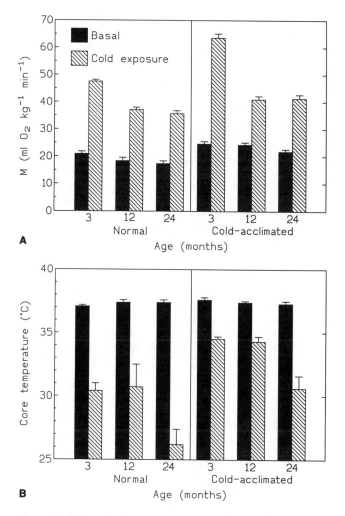

Figure 6.9. Effects of aging on metabolic rate (A) and core temperature (B) for normal and cold-acclimated rats when measured at 28°C (basal) and during cold exposure ($T_a = -10°C$). Metabolic rate was measured during the early phase of cold
exposure; core temperature was measured after 180 min of cold exposure. Data from
Kiang-Ulrich and Horvath (1985).

Facultative thermogenesis. A reduced ability to increase metabolic rate in
response to cold exposure is a predominant thermoregulatory deficit in aged
rats (Table 6.4 and Figure 6.9) and mice (Hoffman-Goetz and Keir, 1984; Talan and Ingram, 1986a). Impaired homeothermy during cold exposure in rats
may be attributable to decay in shivering and nonshivering thermogenic

Table 6.5. *Comparisons of various parameters of BAT in young rats (7 months) and old rats (24 months)*

Parameter	Young	Old
BAT mass		
mg	630 ± 29.5	433 ± 63.9
g/kg body mass	1.72 ± 0.08	1.16 ± 0.16
BAT protein		
mg	94.3 ± 4.3	70.4 ± 4.2
mg/g BAT	154 ± 10.7	145 ± 12.3
BAT mitochondrial protein		
mg recovered	1.74 ± 0.06	1.02 ± 0.12
GDP binding		
nmol/mg mitochondrial protein	0.286 ± 0.03	0.194 ± 0.036
pmol recovered (kg body mass)$^{-0.67}$	9.6 ± 0.8	3.9 ± 1.2

Source: Data from McDonald et al. (1988a).

mechanisms. A marked attenuation in the thermogenic capacity of BAT has been demonstrated in aged rats (Table 6.5); however, there is also a clear sexual dimorphic effect, with males being most severely affected (McDonald et al., 1989). Aged male rats exhibit decreases in BAT mass and GDP binding, whereas aged females have increased BAT mass and stable GDP binding. Moreover, exercise training of aged rats improves their ability to thermoregulate during cold exposure without affecting BAT thermogenic mechanisms (McDonald et al., 1988a). This suggests that some of the failure of cold resistance in rats is attributable to a decline in shivering thermogenesis. On the other hand, exercise training has been shown to have a detrimental impact on cold tolerance in aged mice (Talan and Ingram, 1986b). The potential benefits of exercise in aged humans should be further examined in terms of thermoregulatory control in rodent models of aging.

In view of their attenuated metabolic response to cold exposure, it is not surprising to find that aged rats generally show increased mortality during cold acclimation (Kiang-Ulrich and Horvath, 1985; Owen et al., 1991). On the other hand, another study found better survival rates for 18-month-old rats than for 6-week-old and 6-month-old animals subjected to continuous exposure to an ambient temperature of 4–5°C (Gambert and Barboriak, 1982). The 18-month-old rat is larger than the 6-month-old, which may provide additional insulation against heat loss via a decrease in the surface-area : volume ratio (see Chapter 3). In this case, moderate aging appears beneficial in regard to surviving prolonged cold exposure only because of the increased body mass of the older animals. Obviously, as aging progresses to the point of impaired facultative thermogenesis, the aged rodent is at a decided disadvantage when attempting to thermoregulate in the cold.

Other deficits. Considering the deficits in specific thermoregulatory motor outputs with aging, it is not surprising to find that complex thermoregulatory responses associated with the CTR, exercise, and fever are compromised in aging rodents. At approximately 18 months of age the amplitude of the CTR decreases, a change that is accompanied by reduced cold resistance (Refinetti, Ma, and Satinoff, 1990). Marked impairment in heat dissipation during forced exercise is observed in rats at an age of only 30 weeks, as compared with rats 8 weeks old (Durkot, Francesconi, and Hubbard, 1986); however, this supposed effect of age may be attributable in large part to the increased body mass of the older animals.

Thermal responses to administration of a pyrogen (*Salmonella typhimurium*) are markedly diminished in aged rats kept at 15°C, but not 26°C (Tocco-Bradley, Kluger, and Kauffman, 1985). On the other hand, febrile responses to a yeast injection are relatively unaffected by age in rats (Refinetti et al., 1990). Aged mice exhibit unusual responses to injections of bacterial lipopolysaccharide (LPS) (Habicht, 1981): Young mice, aged 2 and 12 months, become hypothermic, whereas 24-month-old mice become hyperthermic following LPS injection. Clearly, a defect in facultative thermogenesis is likely to have a negative impact on the ability of aged rodents to develop a fever, especially at relatively cool ambient temperatures.

Overall, there is considerable evidence for impairment of thermal homeostasis with aging in rodents. However, many of the data are contradictory. This seems to be the result of inconsistencies in use of techniques (e.g., restraint), genetic strain, gender, and ambient temperature. Moreover, in view of the increased susceptibility of the aged to bacterial infections and altered febrile responses, it is clear that the area of fever and aging deserves more study.

7

Temperature acclimation

Most of the coverage thus far has centered on the responses of rodents' thermoregulatory systems when exposed to heat or cold stress for relatively short durations. However, when thermal stress persists for longer periods (e.g., >24 hr), homeotherms undergo a variety of autonomic and behavioral adaptations that augment their ability to resist adversely warm and cold environments. The purpose of this chapter is to discuss the autonomic and behavioral mechanisms and other adaptations utilized by rodents to efficiently maintain thermal homeostasis when subjected to prolonged periods of heat or cold stress.

7.1. Terminology

One will often find throughout the literature the terms *acclimation, acclimatization,* and *adaptation* used interchangeably to describe the physiological, behavioral, and morphological changes in an organism exposed to an environmental stressor. However, in thermoregulation and other fields dealing with stress, these terms are distinct, and each characterizes a certain set of environmental conditions (IUPS, 1987). *Acclimation* describes the adaptive changes that occur within the lifetime of an organism in response to experimentally induced changes in certain climatic factors. *Acclimatization* is used to describe those changes that occur as a result of inhabiting a particular natural climate. *Adaptation* is a term often used to describe a response in conjunction with acclimation and acclimatization. Conventionally, one can state that an organism adapts, acclimates, or acclimatizes to changes in one or more environmental variables. However, to many biologists, *adaptation* connotes a change in the genotypic makeup of a species. Hence, to be most precise, *genotypic adaptation* describes the genetically fixed changes in an organism that favor its survival under particular environmental conditions

163

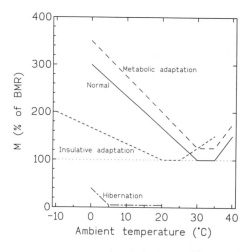

Figure 7.1. Three principal strategies used by homeothermic species to adapt to prolonged cold exposure. *Insulative adaptation* involves an increase in the insulative quality of the fur, resulting in decreased thermal conductance and a lowering of the lower critical temperature, resulting in reduced energy requirements for temperature regulation. *Metabolic adaptation* involves little change in insulative quality, but augmentation of the mechanisms governing resting metabolic rate and nonshivering thermogenesis. During *Hibernation,* body temperature may fall to a level near that of ambient temperature, resulting in an extremely low metabolic rate that is nonetheless capable of increasing in order to prevent freezing of body tissues.

(IUPS, 1987). Relatively few studies in laboratory rodents have addressed the effects of acclimatization and genotypic adaptation on the thermoregulatory system. On the other hand, acclimatization of the thermoregulatory system is frequently studied in humans and other relatively large species. Coverage in this chapter will center primarily on heat and cold acclimation in rodents.

7.2. Acclimation to cold

Three principal strategies have evolved to allow homeotherms to survive prolonged exposure to cold environments: (1) restriction of passive heat loss by enhancement of the insulative quality of the fur, (2) biochemical adaptations of metabolic processes to maintain a higher level of heat production, and (3) torpor and hibernation (Figure 7.1). For nonhibernating species, an insulative strategy is most economical in regard to an animal's energy requirements. Increasing its insulation reduces thermal conductance and its lower critical temperature, thus lowering the species' energy demands for temperature regulation in the cold. Relatively large mammals (cattle, sheep, dogs, foxes, etc.) predominantly display an insulative strategy to adapt to prolonged cold exposure (e.g., Folk, 1974). Because of their relatively large body masses, they have extremely low thermal conductances and are capable of growing thick, insulative coats in the winter. On the other hand, small rodents are limited in their ability to increase their insulation (as discussed later) and thus rely on either a metabolic or hibernation/torpor strategy to survive prolonged exposure to cold. Among the laboratory rodents covered in this book, the golden hamster is the only species capable of hibernation. Various aspects of hibernation in this species are covered elsewhere (see Chapters 2 and 3) and in other publications (Hoffman, 1968; Hart, 1971; Heller, 1979; Lyman, 1990).

7.2.1. Metabolic adaptations

A rodent subjected to continuous cold exposure will undergo an array of adaptations to ensure that its metabolic heat production will balance its increased heat loss. Over the past four decades, research in thermal physiology, covering areas from the molecular level to the whole-animal level, has elucidated a variety of metabolic adaptations proven to be crucial for cold acclimation in rodents (Table 7.1). Most of the biochemical, physiological, and behavioral changes in cold-acclimated rodents may be directly or indirectly related to three major events: (1) replacement of shivering thermogenesis by nonshivering thermogenesis as the principal means of heat production; (2) an elevated basal metabolic rate (BMR); (3) an augmentation in the ability to increase metabolic rate during cold exposure.

Shivering.   During the initial stages of cold exposure, rodents rely on shivering as a primary means of thermogenesis. At first the shivering activity is irregular, but after a few days of cold exposure the shivering becomes nearly continuous (Hart, Heroux, and Depocas, 1956). Shivering is an ephemeral response during acclimation to cold exposure; it is energetically inefficient, it impedes the ability to perform motor tasks, and it disrupts normal sleep patterns (see Chapter 4). The muscular activity from shivering in rats peaks after approximately 1 day of cold exposure, and by 28 days of cold acclimation it declines to a level below that for animals maintained at thermoneutrality (Hart et al., 1956). One study estimated that shivering accounts for less than 10% of the total heat production in rats that are fully acclimated to the cold (Davis et al., 1960). The conversion from shivering to nonshivering thermogenesis during cold acclimation appears to be a common response in other laboratory

Table 7.1. *Summary of adaptive changes in functioning and morphology within various levels of the thermoregulatory system in cold-acclimated laboratory rodents*

Level/effect	References
Central nervous system	
Increases in AVP levels in supraoptic and paraventricular nuclei; reduced febrile response to bacterial pyrogen	Merker et al. (1989)
Altered neuronal responses to changes in skin temperature in dorsal raphe nuclei	Hinckel and Perschel (1987)
Enhanced BAT thermogenic response following stimulation of ventromedial hypothalamus	Thornhill and Halvorson (1990)
Cell membrane (non-BAT sites)	
Enhanced Na^+-dependent ATPase respiration in skeletal muscle, diaphragm, kidney, and liver	Guernsey and Whittow (1981), Horwitz and Eaton (1977), Guernsey and Stevens (1977)
Brown adipose tissue	
Tissue	
Augmented blood flow and oxygen consumption during acute cold exposure and/or administration of sympathomimetics	Foster and Frydman (1978)
Increased density of noradrenergic nerve terminals in BAT	Himms-Hagen (1986)
Cell membrane	
Decrease in density of β-receptors and increase in α_1-receptors; increases in levels of T5'D and lipoprotein lipase	Himms-Hagen (1986)
Increase in ganglioside GM_3 levels	Kuroshima and Ohno (1991)
Mitochondria	
Increases in mitochondrial density, UCP levels, total protein, and cytochrome oxidase activity	Himms-Hagen (1986), Ricquier et al. (1991)
White adipose tissue	
Overall reduction in tissue lipids, but an increase in proportion of unsaturated fatty acids in fat depots	Kodama and Pace (1964)
Reduced size and increased number of epididymal adipocytes; increase in glucagon receptors	Uehara et al. (1986)
Enhanced NE-induced elevation in blood flow to omental depot	Hirata and Nagasaka (1981)
Attenuated rise in plasma FFA during acute cold exposure	Kuroshima et al. (1982)
Skeletal muscle	
Reduced threshold T_c for induction of shivering	Zeisberger and Roth (1988), Brück et al. (1970)
Substitution of nonshivering thermogenesis for shivering as primary heat source during cold exposure	Griggio (1982), Héroux et al. (1959), Pohl (1965)
Increased sensitivity to NE-induced stimulation of oxygen consumption	Shiota and Masumi (1988)

Table 7.1. *(cont.)*

Level/effect	References
Increased myoglobin levels in quadriceps, soleus, diaphragm, and heart	Ohno and Kuroshima (1986)
Endocrine systems	
Adrenal gland	
Increased plasma cortisone levels and augmented release of cortisone in response to fasting	Ohno et al. (1990)
Enhanced metabolic responses to administration of adrenal steroids and epinephrine	Doi and Kuroshima (1984)
Increased sensitivity of visceral organs to epinephrine-induced thermogenesis	Vybíral and Andrews (1979)
Adrenal hypertrophy	Kuroshima et al. (1982)
Pancreas	
Increased glucagon levels in BAT after acute cold exposure	Yahata and Kuroshima (1987)
Increase in plasma glucagon and decrease in insulin	Doi et al. (1982)
Enhanced uptake of glucose by skeletal muscle, heart, and white adipose tissue	Vallerand et al. (1990)
Enhanced thermogenic response to glucagon	Hirata and Nagasaka (1981)
Thyroid	
Increased T_3 and T_4 levels in plasma	Kopecky et al. (1986), Chaffee and Roberts (1971), Fregly (1990)
Increased tissue utilization of T_3 and T_4	Tomasi and Horwitz (1987)
Sympathetic nervous system	
Increased levels of NE and metabolites in blood and urine	Roth et al. (1988, 1990)
Cardiovascular system	
Increased cardiac output at thermoneutrality and during cold exposure	Janský and Hart (1968)
Left ventricular hypertrophy and hypertension	Schechtman et al. (1990)
Increased NE turnover in heart	Jones and Musacchia (1976)
Enhanced NE-induced cardiac output, stroke volume, and coronary blood flow	Hirata and Nagasaka (1981)
Lower T_a threshold for tail vasodilation	Rand et al. (1965)
Renal system	
Polydipsia and diuresis; reduced urine osmolality	Roth et al. (1990)
Metabolism	
Increased BMR; enhanced cold-induced thermogenesis	Banet (1988), Pohl (1965), Chaffee and Roberts (1971)
Increased $\dot{V}O_2$ max	Wang (1981)
Behavior	
Delayed latency to bar-press for radiant heat source	Laties and Weiss (1960)
Reduction in selected T_a	Owen et al. (1991)

Table 7.1. (cont.)

Level/effect	References
Increased food intake and meal size, but reduction in meal frequency	Leung and Horwitz (1976)
Increased food-hoarding behavior	Fantino and Cabanac (1984)
Insulative adaptations	
Reduction in length of tail	Chevillard et al. (1967)
Increase in weight and density of fur	Al-Hilli and Wright (1988a)

rodents, including mouse, gerbil (*Gerbillus*) (Oufara et al., 1987), hamster (Pohl, 1965), and guinea pig (Blatteis, 1976). It should be noted that although shivering is of minor importance, skeletal muscle continues to have a crucial role in thermogenesis during cold acclimation, as discussed later.

BMR.   An elevation in BMR is in itself not needed for cold acclimation, but is a reflection of the immense biochemical changes that take place as nonshivering thermogenesis replaces shivering as the primary source of heat production. Indeed, an elevation in BMR following cold acclimation is clearly a waste of metabolic energy. That is, a cold-acclimated rodent requires an enhanced nonshivering thermogenic capacity only during cold exposure, not at thermoneutrality. Ideally, the most energy-efficient scenario would be to maintain a normal BMR at thermoneutrality but retain the potential to activate nonshivering thermogenesis during cold stress.

An increase in BMR of approximately 20–30% has been repeatedly demonstrated in rats following cold acclimation (Hsieh, 1963; Guernsey and Stevens, 1977; Banet, 1988). Interestingly, whereas the rat has consistently been shown to have an elevated BMR following cold acclimation, studies in other rodent species have been somewhat equivocal. BMR in the mouse and guinea pig apparently is unaffected by cold acclimation (Pohl, 1965; Chaffee and Roberts, 1971; Blatteis, 1976), although these species display marked biochemical adaptations for accelerated nonshivering thermogenesis (Chaffee and Roberts, 1971; Himms-Hagen, 1984, 1990a). Pohl (1965) found no difference in metabolic rates in hamsters acclimated to 6°C and 30°C, but others have found ~28% increases in BMR after acclimation to ambient temperatures of 4–9°C (Adolph and Lawrow, 1951; Tomasi and Horwitz, 1987). It should be added that the golden hamster is one of the most hardy of the rodents when cold-exposed. It has a thick pelt and relatively short limbs and tail, thus aiding in heat conservation. Moreover, compared with the rat, the

Table 7.2. *Tissue oxygen consumption (Qo_2) in pectoral muscle and BMR in rats measured at various time intervals following deacclimation to an ambient temperature of 5°C.*

Parameter	18 hr	7 days	18 days
Tissue Qo_2, [μl O_2 hr^{-1} (mg dry weight)$^{-1}$]			
Control	1.14	1.26	1.68
Cold	1.57	1.41	1.73
%cold/control	37.7	12.1	3.3
BMR (ml O_2 hr^{-1} g$^{-0.75}$)			
Control	6.78	5.12	6.26
Cold	7.75	5.69	6.20
%cold/control	14.3	11.1	−0.9

Source: Data from Guernsey and Whittow (1981).

hamster shows greater metabolic sensitivity to body cooling and markedly better spontaneous rewarming from hypothermia (Adolph and Lawrow, 1951) (see Chapter 5).

The elevations in BMR and the enhanced nonshivering thermogenesis during cold acclimation in rodents are direct consequences of increased activities of catabolic enzymes in BAT (as discussed later), skeletal muscle, and visceral organs. Another important adaptation is the increased activity of the Na-K pump (i.e., Na$^+$-dependent ATPase) in the cell membranes of skeletal muscle, liver, kidney, and other sites (Guernsey and Stevens, 1977; Horwitz and Eaton, 1977; Guernsey and Whittow, 1981). As a related issue, one should note that it is the augmented passive membrane permeability to Na and K that apparently makes mammalian tissues so much more metabolically active than reptilian tissues (Else and Hulbert, 1987). Clearly, the regulation of the Na-K pump is a crucial facet of thermogenesis in cold-acclimated rodents.

At least 1 week of cold acclimation is required before a significant elevation in BMR is observed (Guernsey and Whittow, 1981). In one study, for example, after 6 days of acclimation to 5°C, the BMR was elevated by just 2.6%; however, after 15 days of acclimation, the BMR of the cold-acclimated group was 18.9% above that for control animals (Guernsey and Whittow, 1981). Moreover, when cold-acclimated animals are returned to a thermoneutral environment, the BMR will nonetheless remain elevated for at least 7 days (Table 7.2). The fall in BMR during deacclimation is closely related to the decline in the in vitro metabolic rate of pectoral skeletal muscle. The decline in the oxygen consumption of skeletal muscle during deacclimation from cold is intimately connected with the recovery of Na-K pump activity (Guernsey and Whittow, 1981).

Na-K pump activity in the cell membranes of skeletal muscle and other sites is facilitated by thyroid hormones that are also critical in the control of BMR during cold and heat acclimation (Fregly, 1990). Generally, the blood levels of the thyroid hormones T_3 and T_4 are markedly elevated in rats and other species during cold acclimation (Table 7.1), although a recent study in rats reported only transient elevations in serum concentrations of T_3, T_4, and thyroid-stimulating hormone (TSH) during cold acclimation (Quintanar-Stephano, Quintanar-Stephano, and Castillo-Hernandez, 1991). It is interesting to note that a state of cold acclimation is possibly mimicked by chronic administration of T_3 (1 mg kg^{-1} three times per week). This dosing schedule is reported to lower the ambient temperature of minimal metabolic rate from 28°C to 17.5°C, whereas animals made hypothyroid by carbimazole exhibit an increase from 28°C to 32.5°C (Andrews and Ryan, 1976). Daily injections of T_3 (5 μg/day) lead to better growth in rats acclimating to cold exposure (Hsieh, 1963). The concept of acclimating an animal to cold exposure with pharmacological treatments has obvious benefit in both clinical and agricultural sciences and should be pursued further.

Nonshivering thermogenesis.   A functional sympathetic nervous system is necessary for maintenance of thermal homeostasis during the initial period of cold exposure as well as for development of nonshivering thermogenesis during cold acclimation. During the initial stages of cold exposure, efferent sympathetic nerve terminals release NE, while the adrenal gland increases secretion of epinephrine and corticosteroids. These neural and endocrine responses serve two principal aspects of thermoregulation in the cold: peripheral vasoconstriction of blood flow to reduce heat loss and mobilization of metabolic fuels for thermogenesis (Maickel et al., 1967; Landsberg, Saville, and Young, 1984).

During the initial stages of acute cold exposure, increased blood levels of epinephrine in combination with NE released from sympathetic nerve terminals stimulates lipolysis of triglycerides from fat depots, causing the mobilization of free fatty acids (FFAs) into the circulation. In rats increased releases of palmitic acid (16 : 0) and linoleic acid (18 : 2) are especially prevalent during cold exposure (Figure 7.2). As plasma FFAs increase, the respiratory quotient (RQ) decreases over a period of 4 days, reflecting the preferential metabolism of lipids as a fuel source for thermogenesis (Nakatsuka, Shoji, and Tsuda, 1983) (see Chapter 3). Cold-induced lipolysis is especially prevalent in BAT. In this tissue, lipolysis is achieved through the

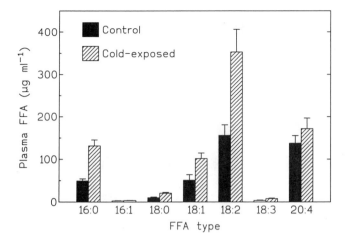

Figure 7.2. Effects of 30 min of acute cold exposure ($T_a = -38°C$) on increases in plasma free fatty acids (FFA) in rats. Data from Ferguson and Shultz (1975).

release of NE from sympathetic nerve terminals and through circulating levels of epinephrine (see Chapter 4). At the beginning of acclimation to cold exposure, lipid stores are reduced in BAT as triglycerides are mobilized into the circulation and also metabolized by BAT. The BAT weight of the hamster is markedly reduced compared with the rat during the first 10 hr of cold exposure (Kopecky et al., 1986). The sympathoadrenal system also plays a key role in pancreatic function by increasing glucagon and suppressing insulin secretion, leading to the stimulation of carbohydrate metabolism in liver and skeletal muscle (e.g., Landsberg et al., 1984) (Table 7.1).

As cold exposure progresses over at least several days, the sympathetic nervous system plays a key role in the development of nonshivering thermogenic mechanisms. Urinary excretions of NE and its metabolites have been found to remain elevated for at least 4 weeks of cold acclimation in several species of rodents. For example, urinary excretion of NE increases by 8.2-fold in hamster (Roth et al., 1990), 3.7-fold in rat (Shum, Johnson, and Flattery, 1969; Papanek, Wood, and Fregly, 1991), and 6.7-fold in guinea pig after 3–4 weeks of acclimation to 4–5°C (Roth, Zeisberger, and Schwandt, 1988). Whereas chronic cold exposure leads to a steady elevation in plasma NE, plasma epinephrine levels in rats increase transiently by ~4-fold after the first day of cold exposure and subside to control levels by 18 days (Papanek et al., 1991). The marked elevation of NE excretion would suggest

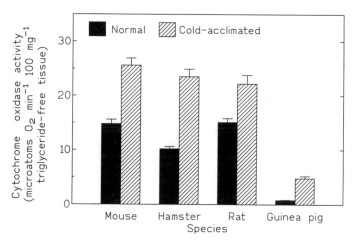

Figure 7.3. Comparison of levels of BAT cytochrome oxidase activity in various laboratory rodents following cold acclimation. Data from Ricquier et al. (1991).

that cold-acclimated rodents are essentially subjected to continuous sympathetic stimulation. This undoubtedly is responsible for many of the changes in various physiological systems during cold acclimation, as summarized in Table 7.1.

The augmented thermogenic capacity of BAT during cold acclimation has been well studied (for further detail of BAT structure and function, see Chapter 4). The sympathetic nerve fibers innervating BAT play key roles in the morphological and functional development of BAT during cold acclimation. Sectioning these nerve fibers attenuates the size and certain functional parameters of BAT in cold-acclimated rats (Park and Himms-Hagen, 1988). Overall, the development of BAT is contingent upon neural and endocrine functions: sympathetic nerve activity to BAT, and a functional thyroid, adrenal, and pancreas. Taken together, a variety of biochemical and morphological adaptations are developed in BAT in cold-acclimated rodents, including increased mitochondrial cytochrome oxidase activity (Figure 7.3), increased concentrations of uncoupling protein (UCP) and triglycerides, increased mitochondrial density, and others; for reviews, see Himms-Hagen (1990a,b) and Nicholls and Locke (1984). It is interesting to note that chronic administration of propranolol, a general antagonist of β-receptors, has no effect on body temperature or BAT development over the time course of cold acclimation in rats (Kortelainen, Huttunen, and Lapinlampi, 1990). Thus, other types of adrenergic receptors (e.g., a_1) and neurohumoral pathways would seem to have major roles in the development of nonshivering thermo-genesis during cold acclimation.

The cardiovascular system is particularly sensitive to the chronic sympa-thetic stimulation associated with cold acclimation. The study of this system is an especially relevant biomedical issue because of the marked incidence of human cardiovascular maladies during the winter months in temperate climates.

Fregly and colleagues have examined the effects of chronic cold exposure on hypertension in rats (Schechtman, Papanek, and Fregly, 1990; Papanek et al., 1991). During chronic exposure to ~5°C, rats exhibit significant eleva-tions in systolic and diastolic pressures, tachycardia, and left ventricular hy-pertrophy. The mean elevations in blood pressure and heart rate in cold-acclimated rats amount to approximately 30 mm Hg and 100 beats per minute, respectively. Moreover, cardiovascular responses to α- and β-adrenergic agonists generally are reduced in cold-acclimated rats (Fregly et al., 1989). Baroreflex sensitivity (i.e., $\triangle$ heart rate/$\triangle$ blood pressure) is also reduced following several weeks of cold acclimation (Papenek et al., 1991). The hypertension that occurs during cold acclimation can be prevented by chronic treatment with captopril, which blocks the conversion of angiotensin I to angiotensin II (Schechtman et al., 1991). The renin-angiotensin system appears to play a key role in the manifestation of hypertension in cold-acclimated rats. Cold-acclimated rodents may provide an excellent experi-mental model for studying hypertension.

Interspecies differences in nonshivering thermogenesis. Some of the most notable interspecies differences in BAT function are seen when comparing the guinea pig with other rodents. For example, the unacclimated guinea pig shows the lowest level of BAT cytochrome oxidase activity, as compared with mouse, hamster, and rat, but the greatest percentage increase in activity fol-lowing cold acclimation (Figure 7.3). An adult guinea pig adapted to normal ambient temperatures has little BAT and apparently requires several months of cold acclimation to achieve the adaptations in BAT morphology and func-tion that are normally seen in the rat after just 3–4 weeks of cold acclimation (Hirvonen, Weaver, and Williams, 1973; Huttunen, Vapaatalo, and Hirvonen, 1975). The patterns of BAT development during cold acclimation in rat and guinea pig are not easily comparable. Kuroshima, Yahata, and Ohno (1991) recently found marked differences in the in vitro sensitivity to NE in BAT from cold-acclimated rats and guinea pigs. BAT thermogenesis during cold acclimation in the guinea pig can be enhanced with regular bouts of forced exercise (Huttunen et al., 1988). This appears to result from the added sympathetic stimulation of BAT during exercise in the cold; however, a similar interaction between exercise and cold exposure in rats yielded equivocal findings (e.g., Richard, Arnold, and LeBlanc, 1986; Shibata and Nagasaka, 1987). It has been suggested that BAT development in the guinea pig may provide the most appropriate rodent model for studies of BAT in humans (Hirvonen et al., 1973). Clearly, more work is needed on the time courses of various functional aspects of BAT and other physiological parameters in different rodent species during cold acclimation.

One test commonly used in rats and other species to assess the degree of cold acclimation is the thermogenic response to a parenteral injection of NE; for reviews, see Janský (1973) and Chaffee and Roberts (1971). A normal rat increases its metabolic rate by approximately 25% at thermoneutrality when given NE (e.g., usually $0.2–0.4$ mg kg^{-1} subcutaneously). A rat fully acclimated to the cold exhibits more than a doubling of its metabolic rate following the same NE administration (also see Figure 4.4). The metabolic response to NE, as measured by percentage increase above the BMR, is directly proportional to the quantity of brown fat as a percentage of body weight (Chaffee and Roberts, 1971). Furthermore, the magnitude of the metabolic response to NE is inversely related to body mass (Hart, 1971; Heldmaier, 1971). The dose of NE needed to achieve a maximal increase in metabolic rate also decreases with increasing body mass.

The duration of cold acclimation required for the development of maximal nonshivering thermogenesis is a meaningful thermoregulatory parameter. It would seem that the quicker an animal can achieve maximal thermogenic capacity, the better its chances of surviving prolonged cold exposure. The thermogenic response to NE in the rat reaches a maximum after approximately 40 days of continuous cold acclimation, whereas 3–8 weeks of deacclimation are needed before NE sensitivity returns to basal levels (Bartunkova, Janský, and Mejsnar, 1971; Yahata and Kuroshima, 1989). It is interesting to note that continuous cold exposure is not needed to augment NE-induced thermogenesis. For example, intermittent cold exposure ($T_a = 5°C$, 2 hr per day for 4 weeks) and continuous cold exposure ($T_a = 5°C$, 4 weeks) elicit similar elevations in the thermogenic sensitivity to NE (Yahata and Kuroshima, 1989). In this regard, intermittent cold exposure ($T_a = 4°C$, 2 hr per day) is also more effective in raising TSH and T_4 levels in rats than is continuous cold exposure (Quintanar-Stephano et al., 1991). It would be helpful to have a better understanding of the interspecies aspects of the time courses of acclimation and deacclimation to cold exposure in rodents.

Partitioning of nonshivering thermogenesis.   An ongoing endeavor by researchers over the past decade has been to identify the major sites for nonshivering thermogenesis. The work of Foster and others using microsphere techniques to measure tissue blood flow has shown that BAT in cold-acclimated rats contributes a major portion of the nonshivering thermogenesis – accounting for up to 29% of the increase in heat production during acute cold exposure (Foster and Frydman, 1979). Surgical removal of 40% of the total BAT mass reduces the thermogenic response to NE by 30% (Rothwell and Stock, 1989). Thus, based on response per unit of weight, BAT is one of the most thermogenic tissues and certainly has been the focus of the majority of the work in nonshivering thermogenesis. It is also clear, however, that non-BAT tissues also must have major roles in heat generation in cold-acclimated rodents.

As mentioned earlier in the discussion of BMR, increased activity of membrane-bound Na-K-ATPase in skeletal muscle, liver, and other non-BAT sites accounts for a large portion of the increased metabolism in cold-acclimated rodents (Table 7.1). These tissues obviously make up a major portion of the body mass of a rodent; hence, a subtle increase in the activity of Na-K-ATPase in these sites, either by an unmasking of inactive enzyme sites or by increased synthesis of new enzymes, could account for a large increase in heat production. For instance, the metabolic activity of the liver can provide a major source of heat, especially following food intake (Guernsey and Stevens, 1977; Adachi, Funahashi, and Ohga, 1991). After 20 days of cold acclimation, the thermogenic response of isolated skeletal muscle to NE increases by more than 50% (Figure 7.4). The NE-induced thermogenesis in skeletal muscle is dependent on Na-K-ATPase activity, providing further support for the existence of sympathetically controlled nonshivering thermogenesis in a non-BAT site. Indeed, it has recently been proposed that the importance of skeletal-muscle thermogenesis has been largely overlooked (Colquhoun and Clark, 1991).

7.2.2. Adaptive changes in CNS during cold acclimation

Considering the adaptations of the motor outputs in cold-acclimated rodents, it would be expected that there would be changes in the processing and central integration of thermoregulatory responses. Brück and colleagues have found that the threshold core temperature for inducing shivering in the guinea pig is lowered following cold acclimation (e.g., Brück and Zeisberger, 1978; Brück and Hinckel, 1990). The shifts in threshold provide

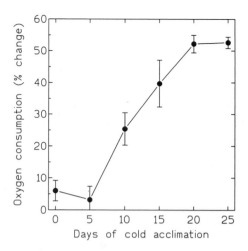

Figure 7.4. Effect of cold acclimation on thermogenic sensitivity of rat skeletal muscle to NE. Data from Shiota and Masumi (1988).

evidence of a change in the neuronal sensitivity in the CNS thermoregulatory centers. It should be noted that shifts in threshold could also result from extra-CNS influences, such as altered thermal sensory function; however, such studies in cold-acclimated rodents have yet to be performed.

Some electrophysiological studies have indicated that cold acclimation can cause direct changes in the CNS. The temperature at which electrical transmission through the hippocampal pyramidal neurons is blocked in the hamster is reduced by 1.9°C after acclimation to 8°C (Thomas, Martin, and Horowitz, 1986). Single-neuron recordings in the raphe nuclei of guinea pig show altered temperature–firing-rate relationships following cold acclimation (Hinckel and Perschel, 1987). A change in the temperature sensitivity of single neurons could mediate altered thermoregulatory thresholds in the cold-acclimated animal. On the other hand, the thermal sensitivity of the preoptic area of the rat, as measured by the change in core temperature during prolonged cooling with stereotaxically implanted thermodes, appears to be unaffected by cold acclimation (Banet and Seguin, 1970). Thus, there could be marked species differences in CNS adaptive responses during cold acclimation (see Section 7.3).

 Evaluation of behavioral variables can also provide information on the changes in the CNS during cold acclimation. Hyperphagia is an example of a behavior that is closely tied in with the metabolic adaptation of cold acclimation. Also, because the regulation of body temperature and the regulation of caloric intake are associated with similar integrative sites in close proximity in the hypothalamus, the effects of prolonged cold or heat stress have provided a useful paradigm for studying the regulation of food intake and body weight (e.g., Magnen, 1983).

To meet the demand for increased heat production during cold exposure, an animal must either consume more food or face a loss of body weight as lipids and carbohydrates and, in the worse case, proteins are metabolized to maintain an elevated heat production. Interestingly, rodents such as the hamster, rat, and guinea pig do not show an immediate increase in food consumption, (i.e., hyperphagia) during the initial stages of cold exposure; a steady-state increase in food consumption is not reached for at least several days (Leung and Horwitz, 1976; Zeisberger, Roth, and Simon, 1988; Roth et al., 1990). This lag in food consumption, coupled with an increased metabolic rate, results in a marked loss in body weight that generally does not stabilize for approximately 7 days of cold exposure. The elevation in food consumption in a cold-exposed rat is quite complex. The increased caloric intake is achieved by a dramatic reduction in the frequency of nocturnal feeding and an increase in the size of the meal eaten (Figure 7.5). Contrary to the lag in increased food consumption during the initial stages of cold exposure, cessation of cold exposure results in immediate reversal of the hyperphagia (Leung and Horwitz, 1976). Food hoarding is also stimulated in cold-acclimated rodents, a beneficial adaptation to minimize sojourns for food in the cold (Fantino and Cabanac, 1984).

7.2.3. Insulative and vasomotor adaptations

Insulation.   Most of the literature on cold acclimation in rodents centers on metabolic adaptations, with little attention paid to adaptations that restrict heat loss. Cold-acclimated rodents do indeed exhibit a variety of insulative and vasomotor adaptations. Although these adaptations are not nearly as elaborate as those seen in larger mammals, they nonetheless aid in lowering heat loss, thus minimizing energy expenditure in the cold (Table 7.1). Of course, growing more fur requires additional energy, but over a long period of time the energy expended in developing insulation is much less than what would have to be used for facultative thermogenesis.

The insulative quality of the fur can be augmented by increasing its thickness and/or density. The former option is limited in relatively small animals such as rodents, because (1) significant increases in the thickness of the fur would impede the animal's movement, and (2) the efficiency of adding

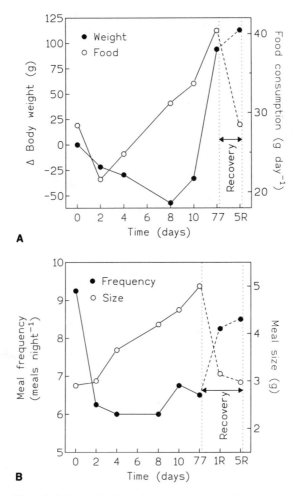

Figure 7.5. Example of complex changes in a rat's feeding behavior during cold acclimation and recovery. A: Note the decrease in food intake during the first 4 days of cold exposure and the subsequent drop in body weight. B: During cold acclimation, nocturnal meal size increases, and meal frequency decreases – an effect quickly reversed after just 1 day of recovery. Data from Leung and Horwitz (1976).

insulation to a cylinder or sphere decreases with decreasing radius. That is, for an inanimate cylinder of radius r, and a given thickness of insulation x, the actual insulation in a plane is calculated by multiplying by the factor $(r/x)\ln[1+(x/r)]$; for a sphere, the insulation is $r/(r+x)$ (Mount, 1971). Thus, the effective insulation of a given thickness is expected to decrease as a function of decreasing body size.

Table 7.3. *Effects of temperature acclimation on various parameters of the fur in mice*

Parameter	Acclimation temperature (°C)		
	33	21	8
Total body hair (mg)	22.8 ± 3.0[a]	41.9 ± 2.0	58.5 ± 4.0
Hair length (mm)			
Abdomen	3.89 ± 0.17	3.88 ± 0.3	6.75 ± 0.44
Interscapular	5.89 ± 0.89	5.78 ± 0.91	5.67 ± 0.37
Dorsocaudal	5.28 ± 0.43	4.06 ± 0.41	7.01 ± 1.45
Hair diameter (μm)			
Abdomen	33.6 ± 3.6	31.8 ± 4.5	41.3 ± 4.5
Interscapular	14.3 ± 2.8	32.2 ± 5.0	32.2 ± 4.8
Dorsocaudal	16.4 ± 1.6	23.4 ± 2.7	33.7 ± 4.0

Source: Data from Al-Hilli and Wright (1988a).
[a] Mean ± S.D.

In spite of the predicted physical limitations, one does see insulative strategies during cold acclimation in some laboratory rodents. For example, the hair thickness and length and the number of hairs per follicle increase with cold acclimation and decrease with heat acclimation in mice (Table 7.3). On the other hand, in one study the insulative qualities for rats were found to be identical in animals acclimated to 6°C or 30°C, whereas the insulative quality of the fur increased markedly in cold-acclimatized rats living in outdoor pens during the winter (Héroux, Depocas, and Hart, 1959). After 65 days of cold acclimation, the weight of the body hair of a rat increases by approximately 60% (Joy and Mayer, 1968). Overall, there is some evidence of an increase in pelt insulation for rats during cold acclimation, but more work in other rodent species is needed.

A reduction in the length of the tail is another common adaptation to limit heat loss during cold acclimation in rats and mice (Chevillard et al., 1963; Chevillard, Bertin, and Cadot, 1967). The tail of the rat accounts for approximately 7% of its total surface area, and at least 20% of its heat loss can occur through the tail when it is vasodilated at thermoneutral ambient temperatures (see Chapter 4). Thus, a shortening of the tail aids in limiting its heat loss in the cold. Moreover, a shorter tail can be shielded from the cold with simple postural adjustments, thus providing easier protection against cold-induced injuries (e.g., frostbite).

Vasomotor.   An initial vasomotor response to cold exposure is peripheral vasoconstriction. Although this response is effective in restricting heat loss, it is apparent that sensory reception and other functions in the cold, vaso-constricted tissues will be compromised. Vasomotor adaptations are essential to increase the skin blood flow sufficiently that sensorimotor reflexes can re-main operative in the cold-acclimated state.

Numerous studies in large mammals and humans have demonstrated in-creased perfusion of cutaneous tissues following prolonged exposure to cold (e.g., Folk, 1974; Hensel, 1981), but relatively little is known regarding the rodents. After nearly 3 months of cold acclimation, the ears of a rat will ex-hibit a slightly warmer surface temperature and a 5.6-fold increase in the number of open capillaries (Héroux, 1959). In vitro measurements of respi-ration in foot skin have shown significant elevations following cold acclima-tion. This elevation in metabolism may reflect the increased mitotic activity of the skin in cold-acclimated animals (Héroux, 1959).

The vasomotor system shows marked adaptations to maintain increased blood flow to the limbs of cold-acclimated rats. Using an anesthetized prep-aration, Brown and Baust (1980) showed that cold acclimation in rats led to enhanced cold-induced vasodilation (CIVD) in a hindlimb exposed to 0°C. In an unacclimated rat, foot skin temperature decreased below 10°C during ex-ternal cooling. The vasoconstrictive response was not localized to the cold-exposed limb, but also occurred in the contralateral, unexposed limb, which implies a general, sympathetically driven vasoconstrictive response. Follow-ing 3–4 weeks of cold acclimation, blood flow to the cold-exposed foot was increased, which prevented the skin temperature from falling below 30°C. After months of cold acclimation, the appendages of a rat were found to be slightly warmer than those of controls during exposure to extremely cold tem-peratures, suggesting an increase in basal blood flow (Banet, 1988). The threshold temperature for tail vasodilation was reduced from 25.9°C to 20.3°C following acclimation to 11°C (Rand et al., 1965). Overall, the shift in threshold temperature for tail vasodilation and the enhanced CIVD to limb cooling indicate an adaptation of the central control of thermoregulatory re-flexes in the cold-acclimated state. Considering these central modifications, along with the localized changes in cutaneous vasomotor functions, it is clear that cold-acclimated rodents are useful models for studying the integrative adaptations of the cardiovascular and thermoregulatory systems.

7.3. Acclimation to heat

Acclimation to heat is perhaps more relevant than cold acclimation in bio-medical research. That is, during the winter, humans don appropriate insu-lation and thus are protected from cold exposure regardless of whether they

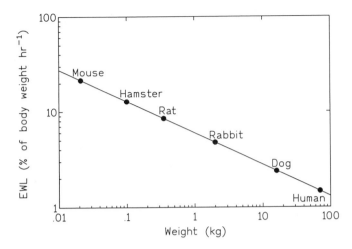

Figure 7.6. Theoretical estimates of water requirements to maintain thermal homeostasis under desert conditions (T_a = 40°C) for mammals of various sizes. Data from Schmidt-Nielsen (1964) and Folk (1974).

are indoors or outdoors. On the other hand, barring individuals who remain in air-conditioned environments throughout the day, acclimatization to heat is generally unavoidable for most humans during the summer months.

Rodents are faced with major physical and physiological limitations when subjected to continuous heat exposure, namely, their relatively small sizes and their inability to sweat or pant. As pointed out in Chapter 3, the small sizes and thus large surface-area : volume ratios for rodents mean not only increased rates of heat loss per gram of body weight during cold exposure but also increased rates of water loss in the heat. Theoretically, under desert conditions (T_a = 40°C) the percentage of body water that must be evaporated per hour in order to maintain a constant body temperature can be shown to decrease logarithmically with increasing body mass (Figure 7.6). This relationship helps to explain the nocturnal behavior of desert rodents: To prevent dehydration they must retreat to their burrows to avoid the heat and low vapor pressure of the daytime, and thus they do most of their foraging at night (Schmidt-Nielsen, 1964).

As ambient temperature increases above thermoneutrality, homeotherms must rely more on evaporation to dissipate body heat (see Chapter 4). Although rodents do not sweat or pant, they can nonetheless increase their evaporative water loss (EWL) by behavioral application of saliva to the skin. Such behavior cannot be maintained continuously for very long, for if it were, the animal would be unable to perform other motor activities. Obviously, increases in this grooming behavior in the heat in the face of limited water availability will rapidly lead to dehydration. It is interesting to note that in rats the regulation of thirst during thermal dehydration is decidedly different from that occurring because of water deprivation (Barney, Williams, and Kuiper, 1991). During water deprivation there is a marked increase in plasma renin, which leads to an increase in angiotensin II, a critical hormone that stimulates thirst and an increase in water intake. On the other hand, thermal dehydration does not appear to activate the renin-angiotensin II system as normally occurs during water deprivation. Overall, rodents are not well suited to maintain a high rate of EWL during prolonged heat exposure. If heat stress cannot be avoided by behavioral means, then heat-acclimated rodents must undergo metabolic and/or insulative adaptations in order to minimize the impact of heat stress on thermal homeostasis and water balance.

7.3.1. Metabolic adaptations

To achieve thermal homeostasis during prolonged heat exposure without severely taxing water balance, rodents have two options: enhance heat loss through insulative or vasomotor adaptations and/or reduce metabolic rate through biochemical modifications. Although there are some insulative adaptations available, such as reducing the insulative quality of the fur, lengthening the tail, and enlarging the surface area of the pinna (Harrison, 1963; Pennycuik, 1969; Al-Hilli and Wright, 1988b), the predominant strategy of heat-acclimated rodents appears to center on adaptations to lower the metabolic rate.

Reductions in metabolic rates in heat-acclimated rodents can arise from two sources: volitional suppression in motor activity, and biochemical adaptations to lower the BMR. Most of the research in this area has been done using the golden hamster and rat. Such a reduction in BMR has been considered to be a reversal of those processes that elevate BMR during cold acclimation, a phenomenon that has been termed "chemical thermosuppression" (Chaffee and Roberts, 1971). During heat acclimation, the oxidative capacities of major heat-producing organs such as the liver, kidney, and BAT are reduced (Table 7.4). For example, the respiratory rate in isolated liver mitochondria from the hamster decreases by 40% after just 10 days of continuous exposure to an ambient temperature of 35°C and recovers to baseline after 4 days of deacclimation to room temperature (Cassuto, 1970). Lowered oxidative capacity is associated with reduced activities of various enzymes involved in oxidative phosphorylation. These biochemical adaptations produce marked reductions in whole-body metabolic rates for heat-acclimated

Table 7.4. *Summary of adaptive changes in function and morphology within various levels of the thermoregulatory systems of heat-acclimated rodents*

Level/effect	References
Central nervous system	
Altered thermoregulatory responses to centrally injected dopamine, serotonin, and PGE$_2$	Ferguson et al. (1984)
Altered responses to centrally injected NE	Young and Dawson (1988)
Decreased hypothalamic levels of NE and serotonin	Tempel and Parks (1982)
Reduced threshold core temperature for salivation and vasodilation of tail	Horowitz and Meiri (1985)
Elevated threshold hypothalamic temperature for EWL and vasodilation of tail	Shido et al. (1991)
Brown adipose tissue	
Reduced weight and protein content	Habara and Kuroshima (1983)
Salivation and evaporative water loss	
Increased weight of salivary glands	Elmer and Ohlin (1970), Horowitz (1976)
Increased concentrations of Na$^+$ and K$^+$ and increased Na$^+$/K$^+$ ratio in saliva	Horowitz et al. (1978)
Increased threshold T_a for EWL	Gwosdow and Besch (1985), Horowitz and Meiri (1985)
Increased EWL	Jones et al. (1976)
Cardiovascular system	
Increase in threshold T_a for tail vasodilation	Rand et al. (1965), Gwosdow and Besch (1985), Horowitz and Meiri (1985)
Increased plasma volume; altered distribution of organ blood flow during heat stress	Horowitz and Givol (1989)
Reduced vascular compliance during hyperthermia	Horowitz et al. (1988)
Reduced ventricular stiffness in isolated heart	Horowitz et al. (1986)
Reduced heart weight	Ray et al. (1968)
Metabolism	
Reduced BMR	Yousef and Johnson (1967), Tal and Sulman (1975), Rousset et al. (1984)
Increase in upper critical T_a	Gwosdow and Besch (1985), Cassuto (1968), Pennycuik (1969)
Endocrine systems	
Adrenal	
Elevated plasma corticosterone levels	Kotby and Johnson (1967), Attah and Besch (1977)
Thyroid	
Increases in plasma levels of T$_3$ and T$_4$	Gwosdow et al. (1985)
Reduced thyroid weight	Ray et al. (1968)
Reduced thyroid production of T$_4$	Rousset et al. 1984)
Pancreas	
Hypoinsulinemia and altered glucose tolerance	Chayoth et al. (1984b)

Table 7.4. *(cont.)*

Level/effect	References
Renal system/water balance	
Reduced clearance of urea, inulin, and PAH	Chayoth et al. (1984b)
Reduced percentage body water; increased turnover of ^3H_2O.	Attah and Besch (1977)
Reduced plasma volume and expanded extracellular volume	Horowitz (1976)
Insulation	
Reduced growth of hair; fewer hairs per follicle	Pennycuik (1969), Al-Hilli and Wright (1988a)
Increased length of tail	Chevillard et al. (1967), Harrison (1963)
Increased surface area of ear pinna	Al-Hilli and Wright (1988b)

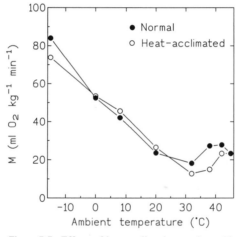

Figure 7.7. Effects of heat acclimation on the ambient temperature–metabolic rate profile of the hamster. Note reduced BMR and increase in upper critical temperature. Data from Cassuto (1968).

rodents and increases in their upper critical temperatures (Figure 7.7) (Yousef and Johnson, 1967; Cassuto, 1968). An increase in the upper critical temperature is adaptive to survival; such a response reduces the metabolic heat load, thus easing the strain on the heat-dissipating effectors.

Not surprisingly, as this so-called chemical thermosuppression develops, the activities of the sympathetic nervous system and hypothalamopituitary-thyroid axis are also suppressed, reflecting the animal's reduced metabolic requirements for thermal homeostasis in the heat. A lowered BMR following heat acclimation is associated with an attenuated metabolic response to ex-

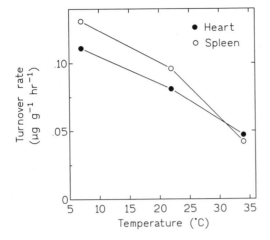

Figure 7.8. Example of the depressant effect of acclimation to warm temperatures on NE turnover in the heart and spleen of the hamster. Data from Jones and Musacchia (1976).

ogenous NE (Arieli and Chinet, 1986). Moreover, NE turnover in the heart and spleen of the hamster acclimated to 34°C is 32–40% of that in animals acclimated to 7°C (Figure 7.8). Urinary excretion of NE is moderately suppressed in hamster and rat when ambient temperature is increased from 22°C to 28–30°C (Shum et al., 1969; Roth et al., 1990). On the other hand, prolonged exposure to extremely warm environments (T_a = 36°C) leads to an elevation in urinary NE, an apparent response to stress that would seem to impede the ability to acclimate to heat exposure (Shum et al., 1969).

A reduction in thyroid function generally parallels the decrease in metabolism in heat-acclimated rodents (Chaffee and Roberts, 1971; Fregly, 1990). Serum T_3 levels in rats acclimated to 34°C are approximately 50% of those seen in animals acclimated to 24°C (Arieli and Chinet, 1986). Furthermore, experimentally induced hypothyroidism enhances heat tolerance, whereas hyperthyroidism or administration of T_4 to animals undergoing heat acclimation attenuates heat tolerance (Fregly, Cook, and Otis, 1963; Cassuto et al., 1970; Chaffee and Roberts, 1971). Because thyroid function is intimately related to body growth and development, it might seem that the reductions in plasma T_3 and T_4 would be major factors in the depression of growth during heat acclimation. However, a study by Rousset et al. (1984) seems to have ruled out a direct effect of reduced thyroid function on depressed growth in heat-acclimated rats.

Reductions in metabolic rate and thyroid activity have profound effects on growth and organ development in heat-acclimated rodents. Food consumption

Table 7.5. *Effects of exposure to ambient temperatures of 22°C or 35°C for 10 weeks on the growth of various organs in rats*

Variable	Acclimation temperature (°C)		Percentage of control
	22	35	
Body weight (g)	389 ± 85[b]	296 ± 56	76
Liver (mg g^{-1})	46.8 ± 11.5	34.7 ± 7.3	74
Kidneys (mg g^{-1})	7.9 ± 1.4	6.6 ± 1.5	83
Lungs (mg g^{-1})	6.7 ± 1.3	6.5 ± 1.5	97
Brain (mg g^{-1})	5.0 ± 0.4	6.2 ± 0.4	124
Heart (mg g^{-1})	3.2 ± 0.8	3.1 ± 0.7	97
Spleen (mg g^{-1})	1.9 ± 0.4	1.8 ± 0.4	95
Adrenals (mg g^{-1})	0.14 ± 0.04	0.14 ± 0.03	100
Thyroid (μg g^{-1})	51.7 ± 11.2	42.1 ± 10.9	81
Pituitary (μg g^{-1})	27.7 ± 8.1	23.3 ± 5.9	84
Testes (mg g^{-1})	8.4 ± 1.6	10.9 ± 1.9	129
Ovaries (mg g^{-1})	0.38 ± 0.1	0.41 ± 0.1	108
Uterus (mg g^{-1})	1.5 ± 0.4	1.8 ± 0.5	120

[a]All organ weights other than those for uterus and ovaries are for male rats.
[b]Mean ± S.D.
Source: Data from Ray et al. (1968).

and growth rate are generally reduced during heat acclimation (Table 7.4). Along with a reduction in body weight, the weights of most organs, excluding brain and reproductive organs, are markedly reduced following heat acclimation in rats (Table 7.5). Some of the most dramatic reductions in organ size are found in the kidney, liver, and heart, which contribute a major portion of an animal's BMR (~35%) (Ray, Roubicek, and Hamidi, 1968; Steffen and Roberts, 1977). Gerbils are unusual in that there is little change in body weight for animals acclimated to 24°C or 34°C; yet there are marked reductions in the sizes and in vitro metabolic activities of major organs, including the heart, liver, and kidneys (Steffen and Roberts, 1977). Moreover, along with reduced organ growth, the functions of the kidneys and liver are also severely compromised in heat-acclimated rodents (Kerr, Squibb, and Frankel, 1975; Chayoth et al., 1984a). This can have a negative impact on the ability to metabolize and excrete drugs and toxic chemicals (see Chapter 9).

7.3.2. Salivary gland adaptations

The development of salivary gland function may be as important to heat acclimation as BAT development is to cold acclimation. The copious production of saliva concomitant with grooming behavior is essentially the only way in which a heat-exposed rodent can actively increase EWL during heat exposure (see Chapter 4). Adaptations that enhance salivary function should benefit acclimation of rodents to hot environments.

During the first 2 days of heat acclimation the submaxillary gland of the rat exhibits dramatic increases in mitotic activity and subsequent hypertrophy (Horowitz and Soskolne, 1978). During this initial phase of heat acclimation there is an approximate doubling of the density of muscarinic receptors in the submaxillary gland, resulting in enhanced sensitivity to cholinergic stimulation (Kloog et al., 1985). As heat acclimation progresses for several weeks, the density of muscarinic receptors returns to basal values. The hypertrophy of the salivary gland can be nearly abolished by sectioning the parasympathetic nerves innervating the salivary gland, whereas sympathectomy has little effect on salivary function during heat acclimation (Elmer and Ohlin, 1970). The increased mitotic activity of the salivary gland subsides after approximately 5 days of heat acclimation, but hypertrophy is maintained for at least several weeks (Elmer and Ohlin, 1970; Horowitz, 1976). Paradoxically, in spite of the increased development of the salivary gland and enhanced ability to increase EWL during heat exposure (Jones, Musacchia, and Temple, 1976; Gwosdow and Besch, 1985), the production of saliva in such animals has been found to be less than that in controls (Horowitz et al., 1983; Horowitz and Meiri, 1985). Furthermore, cholinergic stimulation of salivary flow by administration of pilocarpine is also reduced in heat-acclimated rats (Kloog et al., 1985). Clearly, further study is needed regarding the relationship between salivary gland development and function in other rodent species during heat acclimation.

7.3.3. Adaptive changes in CNS during heat acclimation

There is substantial evidence from several lines of study that the CNS undergoes marked alterations over the time course of heat acclimation in rodents. Heat-acclimated rats consistently show altered thermoregulatory sensitivity to localized injections of various neuromodulating agents into the POAH, including NE, 5-HT, and PGE_2 (Ferguson et al., 1984; Christman and Gisolfi, 1985; Young and Dawson, 1988) (Figure 7.9). Interestingly, the thermal sensitivity of trigeminal neurons receiving inputs from facial warm and cold receptors (see Chapter 2) is unchanged in rats reared at ambient temperatures of 20°C and 30°C (Dawson et al., 1982). This implies that the thermal sensory circuits are essentially "hard-wired" and cannot be influenced

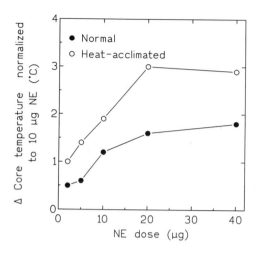

Figure 7.9. Change in POAH sensitivity to microinjection of NE in the heat-acclimated rat. Data from Christman and Gisolfi (1985).

by thermal acclimation, a conclusion confirmed in studies of warm- and cold-acclimated cats (Hensel and Schäfer, 1982).

The neurochemical changes in the CNS are most likely associated with shifts in sensitivity of thermoregulatory motor outputs and may possibly explain changes in the set-point. Elevating the set-point during heat acclimation would be adaptive in a variety of ways. When the regulated core temperature is raised, the error signal on the thermoregulatory controller (i.e., $T_c - T_{set}$) is reduced, resulting in a lower demand for heat-dissipating motor outputs (see Figure 1.3). The perceived heat load will also be expected to decrease, which will lessen the psychological stress of the elevated temperature. Apparent elevations in the set-point have been documented in nonrodent species such as the cat and camel when exposed to combinations of heat and dehydration. In these species, the core temperature is allowed to rise much higher than the normal threshold before there is an increase in EWL (Schmidt-Nielsen et al., 1957; Doris and Baker, 1981). Thus, by elevating the set-point, thermoregulatory control is compromised, but precious body water is conserved.

Can similar responses be operative in heat-acclimated rodents? The evidence for an elevation in set-point following heat acclimation in rodents is species-dependent and equivocal. Brück et al. (1970) first reported that heat-acclimated guinea pigs had an elevated set-point for regulating body temperature. This was apparent from the increased threshold core temperature for activating thermal polypnea and maintenance of an elevated core temperature

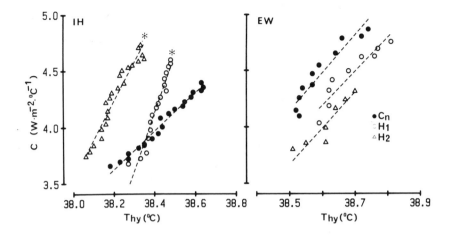

Figure 7.10. Effects of heat acclimation on the sensitivity of thermal conductance to increases in hypothalamic temperature (T_{hy}) induced through internal heating (IH) and external warming (EW) in rat. (acclimation groups: C_n, 24°C; H_1, 29.4°C; H_2, 32.8°C). Reprinted from Shido and Nagasaka (1990a) with permission from The American Physiological Society.

when the heat-acclimated animals were returned to a relatively cool environmental temperature.

The evidence for a change in set-point in rats following heat acclimation is somewhat conflicting. Unanesthetized, heat-acclimated rats exhibit elevated threshold ambient, core, and hypothalamic temperatures for increasing EWL and tail skin blood flow (Shido et al., 1991), whereas another study found reduced threshold core temperatures for salivation and vasodilation (Horowitz and Meiri, 1985). Contrary to the guinea pig, the heat-acclimated rat (5 hr per day; $T_a = 33°C$ or 36°C) has a lower core temperature when it is returned to a relatively cool temperature of 24°C (Shido and Nagasaka, 1990a). On the other hand, rats continuously exposed to 32.4°C for 10 days had higher core temperatures for 2 days of deacclimation at 24°C (Shido et al., 1989). It generally requires 5–10 days of continuous heat acclimation to observe distinct changes in threshold body temperatures for activating thermoregulatory effectors (Shido et al., 1991). To complicate matters, the sensitivities of peripheral vasodilation, thermal conductance, and EWL responses to changes in core temperature vary depending on whether the body heat is internal or external and whether the heat acclimation is continuous or intermittent (Shido, Yoneda, and Nagasaka, 1989; Shido and Nagasaka, 1990a,b) (Figure 7.10). Behavioral thermoregulatory studies are needed to further delineate the occurrence of changes in the set-point during heat acclimation. That is, a true elevation in the set-point should be associated with a preference for warmer environmental temperatures, as well as an elevated threshold for activating heat-dissipating autonomic responses.

8

Gender and intraspecies differences

When reproductive status is not an issue, physiologists generally study the characteristics of male members of a species. Seldom does one find thorough analyses of the influence of either gender or genetic strain (i.e., intraspecies variation) on thermoregulation or other regulatory systems. Moreover, one occasionally finds that males and females have been used interchangeably in a study with no regard to the possibility of sexual differences. Likewise, the genetic strain often is not considered as a significant factor in a study. Such omission is neither intentional nor the result of ignorance. Generally, as is the case in thermal physiology and other fields, there is so little information on the impact of gender and genetic strain that it is impossible to consider these factors in interpreting the data. Hence, this chapter is intended to provide a review of the effects of gender and genetic strain on thermoregulatory responses in rodents. It is hoped that an analysis of the meager data base in this area will spur more research in the areas of gender and strain differences in thermal physiology.

8.1. Gender differences

Considering the obvious morphological and physiological variations of male and female rodents, one should not be surprised to find gender-specific thermoregulatory differences. The action of steroid sex hormones (i.e., androgens and estrogens) can impact in various ways on thermoregulatory processes. It was discussed earlier (Chapter 6) how females undergo marked changes in thermoregulatory sensitivity during pregnancy and lactation. However, as will be discussed later, relatively little is known regarding the thermoregulatory characteristics of virgin and nonpregnant rodents as compared with the males of a species.

191

8.1.1. Thermoregulation and the estrous cycle

Whereas testicular function in male rodents is relatively constant from one day to the next, females exhibit a distinct 4–5 day reproductive cycle under control of the hypothalamopituitary-ovarian axis: proestrus, estrus, metestrus, and diestrus. Effects on body temperature are usually apparent during the nocturnal phase of estrus, the time of ovulation that coincides with peak elevations in gonadal and adenohypophyseal hormones. The body temperature of the rat is approximately 0.5°C higher at night during estrus, but appears to remain unchanged during the daytime throughout the reproductive cycle (Yochim and Spencer, 1976; Yanase, Tanaka, and Nakayama, 1989). This information can be useful to those wishing to compare the responses of male and female rats in which thermoregulation may be an issue. It would seem that parameters such as metabolism and body temperature can be compared during the daytime in male and female rats regardless of the status of the female's estrous cycle.

The thermoregulatory pattern of the female rat during the estrous cycle is clearly modulated by changes in motor activity. Running-wheel activity and metabolic rate in rats show marked elevations during the night of proestrus (Anantharaman-Barr and Decombaz, 1989). A recent study in rats demonstrated that body temperature was highest during proestrus and correlated with peak elevations in running-wheel activity (Kent, Hurd, and Satinoff, 1991). Moreover, when rats are prevented from running, the temperature changes due to the estrous cycle are nearly abolished, suggesting that motor activity has a major influence on temperature oscillations during the estrous cycle. The estrous temperature cycle could indeed represent a paradigm of motor activity as a true thermoregulatory effector (see Chapter 4). That is, the heat generated by motor activity generally is not viewed as an effector response, but estrus-induced motor activity appears to be a clear exception.

There has been relatively little work on the control of other thermoregulatory effectors during the estrous cycle. The threshold rectal temperature for stimulating tail vasodilation is higher during estrus, suggesting that the set-point for body temperature is elevated during this stage of the reproductive cycle (Yanase et al., 1989). Interestingly, the golden hamster differs from the rat in that the normal elevation in metabolic rate at night is suppressed during estrus, rather than accentuated as has been observed in the rat (Schneider, Palmer, and Wade, 1986). Many unresolved issues need to be addressed regarding comparative differences in thermoregulation during the estrous cycle.

8.1.2. Metabolic rate and cold resistance

Resting metabolic rate and core temperature generally are reported to be higher in female rats (Castella and Alemany, 1985; Fujii and Ohtaki, 1985), although in a very early but thorough study of rats it was found that males had higher metabolic rates when maintained at thermoneutral temperatures (Benedict and MacLeod, 1929). Furthermore, metabolic rates for male and female rats were found to be similar at ambient temperatures of 5–25°C, but core temperatures for the females were significantly higher at both ambient temperatures (McDonald et al., 1989). Sexual dimorphic differences in metabolic rates have also been reported for golden hamsters, with males having 8–15% higher rates than females, depending on the time of year (Robinson, 1968). Contrarily, the female hamster has been shown to have a consistently elevated body temperature measured over a 24-hr period (Chaudhry et al., 1958). It would seem that the findings of metabolic-rate differences in male and female rodents are equivocal and deserve further study.

Gender differences in thermoregulation can be more easily detected when the animals are exposed to adversely hot or cold environments. During acute cold exposure, male rats exhibit greater elevations in shivering activity and metabolic rate than do females (Figure 8.1). Even though the mature female is smaller than the male rat, core temperatures are similar for the sexes during cold exposure, which suggests that insulation or other factors allow the female to thermoregulate more efficiently (Doi and Kuroshima, 1982b). The female rat also has an attenuated thermogenic response to NE administration that can be reversed by ovariectomy (Doi and Kuroshima, 1982b). Adult male and female rats exposed continuously to an ambient temperature of 2°C experience incredible differences in survival (Zarrow and Denison, 1956). The time for 50% survival in that cold environment was 8 days for males and 90 days for females; however, castrating the males resulted in lengthening their cold survival to match that of the females. It is clear from these data that ovarian and testicular functions modulate thermoregulation in the cold.

Studies utilizing administration of steroidal hormones have helped to elucidate the mechanisms of sexual dimorphic differences in thermoregulation. For example, whereas ovariectomy lowers the body temperature (Yochim and Spencer, 1976), both progesterone and estradiol treatments will raise metabolic rate and body temperature in ovariectomized animals (Marrone, Gentry, and Wade, 1976; Laudenslager et al., 1980). The hyperthermic effect of progesterone is effective in the intact rat and ovariectomized rat, but ineffective after animals have been hypophysectomized (Freeman et al., 1970). Furthermore, chronic treatment with estradiol has little effect on BAT

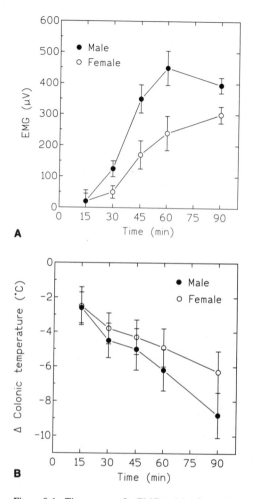

Figure 8.1. Time courses for EMG activity (i.e., shivering) of neck muscles (A) and decreases in colonic temperature (B) in 65-day-old male and female rats during exposure to an ambient temperature of 5°C for 90 min. Rats were urethane-sedated. Data from Doi and Kuroshima (1982b).

function at thermoneutrality, but exerts suppressive effects on various parameters of BAT thermogenesis in cold-acclimated rats, including BAT protein content, cytochrome oxidase activity, and GDP binding (Puerta et al., 1990). In contrast, treatment with progesterone has little effect on these same parameters in normal and cold-acclimated rats (Nava, Ablenda, and Puerta, 1990). Estrogen treatment has also been found to affect behavioral thermoregulation by stimulating operant heat reinforcements in ovariectomized, cold-exposed rats (Wilkinson, Carlisle, and Reynolds, 1980). Overall, there is a complex interaction between ovarian function and thermoregulation during short-term cold exposure and cold acclimation that requires further study.

8.1.3. Heat resistance and fever

Sexual dimorphic studies on thermoregulation in the heat have been few and somewhat controversial. The most thoroughly studied aspect pertains to the role of the scrotum. In addition to its reproduction functions, the scrotum serves as both a heat sensor and dissipator. During exposure to heat stress, the scrotal sac engorges with blood to facilitate dry heat loss. The scrotum of the rat is extremely sensitive to temperature elevations and is well innervated with thermoreceptors (see Chapter 2). Heating the scrotum of the rat activates thermointegrative neurons in the CNS and also elicits vasomotor responses in the tail (Neya and Pierau, 1976; Ishikawa et al., 1984). Male rats are more active in the heat and spend more time grooming saliva onto the base of the tail and scrotum, which augments evaporative heat loss (Hainsworth, 1967).

During exercise, the male rat has also been found to increase the blood flow to its tail at a lower core temperature than the female, consequently maintaining its core at a lower temperature during exercise (Thompson and Stevenson, 1965). On the other hand, Furuyama (1982) reported a significantly longer survival time for female Sprague-Dawley rats during exposure to acute heat stress (T_a = 42.5°C). It appears that little attention has been focused on the possible changes in heat tolerance in female rodents during the estrous cycle. It would seem that the oscillations in steroidal and hypothalamic hormones, body temperature, and motor activity would affect key facets of temperature regulation in the heat.

Sexual differences can also be found in regard to the onset of and recovery from fever. Ford and Klugman (1980) found that intraperitoneal administration of bacterial endotoxin led to a typical febrile elevation in body temperature in male rats, but a fall in temperature in females. The hypothermia in females was associated with a rise in skin temperature, suggesting that altered cutaneous vasomotor responses could explain the lack of fever in the females.

Testicular function has also been found to be important in the activation of thermolytic motor outputs (Pittman et al., 1988). Castration leads to neurochemical changes in the CNS and specifically results in marked reductions in arginine vasopressin (AVP) levels in the ventral-septal area. Because AVP is crucial in the lowering of body temperature during fever (see Chapter 2), it

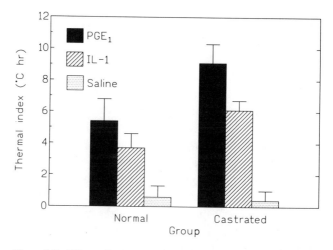

Figure 8.2. Effects of castration on the thermal index in male rats following intraventricular injection of PGE_1, IL-1, or saline. The thermal index is an integration of the change in core temperature with time, yielding dimensions of degrees Celsius times hours. Data from Pittman et al. (1988).

was reasoned that castration would also affect heat dissipation. Indeed, 4 months after castration, male rats exhibit longer fevers following injections of PGE_1 and IL-1 and also have elevated body temperatures during exposure to a warm environment (Figure 8.2). Interestingly, thermoregulation in a cold environment appears to be unaffected by castration. Thus, it seems likely that events that affect testicular function could also have profound effects on thermoregulation. In this regard, it was noted earlier (Chapter 6) that prolonged exposure to warm ambient temperatures leads to a regression in testicular function, including reduced serum levels of testosterone. It seems possible that heat-induced testicular dysfunction could lead to thermoreguatory deficits during heat acclimation.

8.2. Intraspecies differences

Selective breeding of laboratory rodents has led to the development of countless genetic strains to serve the needs of researchers in a variety of disciplines. There seems to have been little research devoted to developing inbred strains for specific thermoregulatory characteristics. However, it turns out that popularly used stains of the mouse and rat, as well as breeds with specific genetic defects, possess distinct thermoregulatory characteristics. It is of obvious importance to understand these intraspecies differences be-

Table 8.1. *Comparison of hypothermic effects of morphine in two genetic strains of mice of similar body weights*

Parameter	Strain	
	DBA/2	C57BL/6
Weight (g)	21.3 ± 0.5	21.8 ± 0.3
Basal T_c (°C)	37.1 ± 0.2	37.0 ± 0.2
$\triangle T_c$ (°C), 30 min after:		
Saline	1.40 ± 0.45	1.25 ± 0.39
Morphine[a]	−3.65 ± 0.48	−0.95 ± 0.36

[a]20 mg kg^{-1} subcutaneously.
Source: Data from Muraki and Kato (1987).

cause the thermoregulatory response can have such a profound influence on the animal's sensitivity to nutritional, pharmacological, and toxicological agents.

Some studies have explored the effects of genetic strain on thermoregulatory responses to various chemical agents. The thermoregulatory effects of morphine have been found to be dependent on the genetic strain (Rosow et al., 1980; Muraki and Kato, 1987). The hypothermic effect of morphine is more pronounced in mice of the DBA/2 strain than in the C57BL/6 strain, even though the two strains are similar in body mass (Table 8.1). The C57BL/6 strain also has an unusually low amplitude of its circadian body temperature rhythm compared with other mouse strains (Connolly and Lynch, 1981). Russel and Overstreet have developed two genetic strains from the Sprague-Dawley line, one sensitive and one resistant to the hypothermic effects of cholinergic agonists and anticholinesterase inhibiting agents (Overstreet et al., 1988). Body temperature in rodents is often used as a sensitive endpoint in the screening of drugs and other chemical agents (see Chapter 9). Thus, pursuing the underlying mechanisms of genetic sensitivity to chemical-induced alterations in thermoregulation should be a viable area of research.

There are marked behavior and autonomic thermoregulatory differences between albino and pigmented rats that are just beginning to be assessed. Pigmented rats such as those of the Long-Evans (LE) strain generally are more active than rats of the albino Fischer 344 (F344) strain and have higher core temperatures under standard housing conditions (see Figure 5.1). When first placed in a temperature gradient, rats of the LE strain will prefer a much cooler ambient temperature than will rats of the F344 and Sprague-Dawley (SD) strains (Gordon, 1987). However, over a period of 6 hr, all strains will begin to select similar temperatures (see Chapter 4).

Table 8.2. *Interscapular BAT variables in LE and SD rats acclimated to 21°C*

Variable	Long-Evans	Sprague-Dawley
body weight (g)	478	474
BAT weight		
mg	358.8	174.6
mg/g body mass	0.75	0.36
Protein (mg)	18.7	11.5
Mitochondrial protein (mg/organ)	35	28
Mitochondrial GDP binding (nmol/mg protein)	0.24	0.15
Mitochondrial UCP protein (μg/mg mitochondrial protein)	44	26

Source: Data from Thornhill and Halvorson (1992).

Various differences in autonomic thermoregulatory responses have been reported for rats. The SD strain has been shown to have better heat tolerance than the LE strain (Furuyama, 1982), but the SD strain is markedly less cold-tolerant than the F344 strain (Kiang-Ulrich and Horvath, 1984b). The LE strain shows a relatively small increase in EWL when exposed to an ambient temperature of 34°C, as compared with the SD strain; the F344 rat likewise shows little rise in EWL. It appears that the F344 rat is one of the better-adapted rats for maintaining a stable body temperature with rising ambient temperature (Gordon, 1987).

BAT development in rats fed normal and "cafeteria" diets has been shown to vary as a function of genetic strain (Rothwell, Saville, and Stock, 1982). Thornhill and Halvorson (1992) recently found that BAT structure and function are extremely well developed in the LE rat as compared with the SD strain. Indeed, it was found that BAT mass for LE rats maintained at room temperature was approximately equal to that for cold-acclimated SD rats (Table 8.2). Overall, quantitative differences in thermoregulatory function are common in the most frequently used strains of laboratory rats. Researchers should be cognizant of these differences when selecting a given rat strain as an experimental model in any study in which thermoregulatory sensitivity may be a critical issue.

8.2.1. Genetic deficiencies in thermoregulation

Obese rat and mouse. The development of genetic rodent models for the study of obesity led to the fortuitous discovery of a correlation between thermoregulatory dysfunction and obesity. This link has provided a useful paradigm for studying the integrative control of basal and facultative thermogenesis and caloric balance. In view of their overlapping control in the hypothalamus, it is clear that thermoregulation and caloric balance are intimately related. Hence, it should not be surprising to find simultaneous defects in these regulatory systems.

There are two principal genetic lines that display marked deficiencies in lipid metabolism and control of body weight: the *ob/ob* mouse and the Zucker *fa/fa* rat; for reviews, see Himms-Hagen (1989, 1990a). In both mouse and rat, the *fa* and *ob* alleles are recessive. When the alleles are homozygous, the animal's body weight may be double that of its lean littermates (i.e., *ob/+* or *fa/?*), which is basically attributable to hyperphagia and/or a reduced metabolic rate. Moreover, the *ob/ob* mouse and *fa/fa* rat show a marked inability to thermoregulate when exposed to cold temperatures. The onset of obesity is associated with a reduced energy expenditure, whereas energy intake may remain relatively constant. Interestingly, similar deficiencies in weight control and thermoregulation are seen in mice exposed neonatally to an acute dose of glutamate and in adult mice acutely treated with gold thioglucose (Himms-Hagen, 1989, 1990a).

Facultative thermogenesis in the *ob/ob* mouse is defective. At ambient temperatures below thermoneutrality, *ob/ob* mice have a reduced metabolic rate and a core temperature that is 1–2°C below that of lean animals (Kaplan and Leveille, 1974; Ohtake, Bray, and Azukizawa, 1977; Trayhurn and James, 1978). During cold acclimation (14°C), the development of BAT thermogenesis, as indicated by mitochondrial GDP binding and increased T5'D activity, is markedly reduced in *ob/ob* mice (Kates and Himms-Hagen, 1990). These biochemical changes in BAT are remarkably dependent on circulating levels of corticosteroids, because adrenalectomy prior to cold exposure improves the *ob/ob* mouse's BAT thermogenic responsiveness. Not surprisingly, the *ob/ob* mouse is not notably cold tolerant and will succumb quickly to hypothermia during exposure to 4°C (Trayhurn and James, 1978). Interestingly, both young and adult *ob/ob* mice display reduced motor activity, which also contributes to the lower metabolic rate and increased body weight (Dauncy and Brown, 1987).

Why do *ob/ob* mice maintain a lower core temperature? Is the hypothermia due to a defect in thermogenesis, or is it mediated centrally via a reduction in the set-point? Measurements of behavioral thermoregulatory responses should aid in clarifying this issue; however, the results to date are somewhat controversial. Two temperature-gradient studies have reported an increase and a decrease in selected T_a for *ob/ob* mice relative to their lean littermates (Carlisle and Dubuc, 1984; Wilson and Sinha, 1985). When provided with an operant mechanism to activate a radiant heat source in the cold ($T_a = 0°C$),

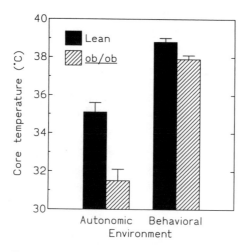

Figure 8.3. Comparison of the effectiveness of autonomic and behavioral thermoregulatory responses in the *ob/ob* mouse. Core temperature was measured in *ob/ob* mice
and lean mice after 30 min of exposure to 0°C. In the behavioral experiment, core
temperature was measured after 30 min at 5°C while the animals could bar-press for
a radiant heat source. Data from Dauncey (1984).

ob/ob mice will select more heat reinforcements and keep their core temperature equal to that of the lean controls (Figure 8.3). Without the radiant heat
source, core temperature for the *ob/ob* mice declines at more than double the
rate for the lean mice. Recent studies in newborn *ob/ob* mice indicate that a
preference for warmer ambient temperatures can be detected as early as 12
days of age (Wilson, Currie, and Gilson, 1991). Moreover, newborn *ob/ob*
mice in a temperature gradient will select an environment that will raise their
core temperature to near that of their lean littermates, a finding similar to
that in adult *ob/ob* mice discussed earlier. Thus, whereas autonomic defi-
ciency of thermoregulation is clear in the *ob/ob* genotype, it seems that behavioral thermoregulatory responses are fully operative and are used to
regulate core temperature.

The *fa/fa* rat is similar to the *ob/ob* mouse in that its facultative thermogenesis has generally been found to be impaired. Thermoregulatory deviations in the *fa/fa* rat can be seen as early as postnatal day 6 (Figure 8.4). In
àdult *fa/fa* rats, metabolic rate normalized to the fat-free body mass is actually higher than in lean animals (Demes et al., 1991). It has been concluded
that *fa/fa* rats have an essentially normal thermogenic response, although the
metabolic rate per unit of surface area is nonetheless lower in cold-exposed
fa/fa rats.

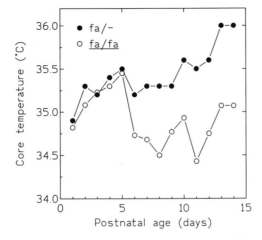

Figure 8.4. Core-temperature differences in Zucker (*fa/fa*) and lean (*fa/−*) rats can be detected as early as 6 days of age. Core temperature was measured in huddling neonates maintained at an ambient temperature of 25°C. Data from Schmidt et al. (1984).

The *fa/fa* thermoregulatory system is functional, but operates at a lower core temperature. For example, adult *fa/fa* rats show an increase in facultative thermogenesis with decreasing ambient temperature while maintaining a lower core temperature (Armitage et al., 1984). Furthermore, the reduced core temperature of *fa/fa* rats is remarkably constant (~1°C) over ambient temperatures of 5–30°C. This implies that the set-point is reduced in the *fa/fa* rat, similar to that in the *ob/ob* mouse. However, a thorough analysis of the behavioral thermoregulatory responses of the *fa/fa* rat must be performed before firm conclusions can be reached regarding the role of the set-point.

Spontaneously hypertensive rat. The spontaneously hypertensive rat (SHR) strain has become a valuable experimental model for studying the impact of stress on various physiological systems, especially during the development of hypertension and related cardiovascular maladies. Compared with its normotensive Wistar littermates, the SHR has elevations in peripheral vascular resistance and in mean arterial blood pressure (i.e., usually the spontaneously hypertensive state is defined as a mean blood pressure exceeding 150 mm Hg). The cardiovascular abnormalities of the SHR generally are attributed to an overactive sympathoadrenal system; for a history of development, see Okamoto and Aoki (1963).

Considering the importance of sympathetic discharge for the maintenance of thermoregulatory tone, it was not surprising in early studies to find thermoregulatory abnormalities in the SHR strain, especially in the control of body temperature during heat stress. For example, heat tolerance, as measured by the time to reach lethal body temperature, is reduced by over 50% in the SHR strain (Wright et al., 1977a). The SHR shows an atypical type II heating pattern, rather than type III (see Figure 5.4), and a reduced ability to increase its EWL during acute heat stress (Wright et al., 1977a). Heart rates during acute heat stress are nearly equal in SHR and normotensive rats, whereas venous blood pressure and abdominal aortic blood flow are markedly reduced in the SHR strain (Wright, Knecht, and Toraason, 1978b). These cardiovascular abnormalities likely contribute to the deficient heat resistance observed in the SHR strain.

The SHR phenotype is also linked with various autonomic and behavioral thermoregulatory deficiencies. Many researchers have reported elevated metabolic rates in the SHR strain at ambient temperatures both within and outside the thermoneutral zone (Wright et al., 1978a; Collins et al., 1987). One operant thermoregulatory study found that the SHR strain preferred warmer temperatures than did normotensive controls (Wilson and Wilson, 1978). This suggests that the set-point for body temperature may be elevated in the SHR phenotype.

The enhanced activity of the sympathetic nervous system and the elevated metabolic rate are key factors in contributing to the elevated core temperatures in the SHR strain (Wright et al., 1978a; Hajos and Engberg, 1986; Collins et al., 1987). However, Morley et al. (1990) noted that in previous studies of the SHR strain, core temperatures had been measured in animals that had been either restrained or subjected to substantial handling stress. Considering that the SHR strain has an exaggerated sympathetic response to stress, it seemed possible that the higher core temperature was an indirect consequence of stress. Indeed, it was found that core temperature in the SHR strain measured by biotelemetry was indistinguishable from that for normotensive controls (Morley et al., 1990). Thus, future studies with the SHR strain should take into consideration the effects of restraint and handling on core temperature and related thermoregulatory responses.

Dystrophic hamster.   The dystrophic golden hamster is a useful model for studying muscular dystrophy and related neuromuscular diseases. The dystrophic phenotype is transmitted with a single recessive gene, and in the homozygous state it is associated with degeneration of skeletal muscle and cardiac muscle (Homburger, 1979). Horwitz and Hanes (1974) first noted the impaired ability of the dystrophic hamster to increase its metabolic rate following catecholaminergic stimulation. This deficit in facultative thermogenesis was originally attributed to the degenerated condition of the hamster's skeletal muscle. However, it was later discovered that the weight, protein content, and cytochrome oxidase activity of BAT were markedly reduced in the dystrophic hamster (Himms-Hagen and Gwilliam, 1980). Thus, the impaired metabolic response to catecholamines is attributed to both an underdeveloped BAT and a dysfunction in skeletal muscle thermogenesis.

The developing dystrophic hamster begins to show reduced body temperatures at an age of 30 days, and by 160 days its body temperature is 1.7°C below that of normal hamsters (Desautels, Dulos, and Thornhill, 1985). Interestingly, the dystrophic hamster can acclimate to 4°C while maintaining a core temperature slightly below that for animals kept at 21°C. Core temperatures are also slightly reduced in dystrophic hamsters kept at warm ambient temperatures.

The food intake of the dystrophic hamster is markedly below that for control animals, which may be a reflection of the poorly developed BAT and subsequent reduction in dietary-induced thermogenesis (see Chapter 4). Metabolic rates in dystrophic and normal hamsters are generally similar (Desautels et al., 1985). Thus, the dystrophic hamster features poorly developed BAT, impaired food consumption, and lower body temperature, although its metabolic rate appears unaffected. An interesting hypothesis is that the lower core temperature may actually benefit the dystrophic hamster's altered protein metabolism (Desautels et al., 1985). Measurements of the dystrophic hamster's behavioral thermoregulation and peripheral vasomotor responses should provide invaluable data for understanding its altered regulation of core temperature.

9

Thermoregulation during chemical toxicity, physical trauma, and other adverse environmental conditions

Once thermoregulatory reflexes are fully developed in postweaning rodents, core temperatures are maintained within relatively narrow limits throughout the animals' lives. This exquisite regulation is compromised only during environmental heat and cold stress, exercise, fever, and adverse environmental conditions. Much of this book has focused on the characteristics of rodents' thermoregulatory systems when subjected to thermal stress. In the existing literature, little attention has been paid to the study of thermoregulation when subjected to other adverse environmental conditions. In this context, "adverse" refers to chemical, pathological, traumatic, and other stressors that may exert effects on the behavior and/or autonomic physiology of an organism. This chapter is an attempt to show not only how rodents' thermoregulatory systems are especially susceptible to such stimuli but also how their regulatory characteristics are altered in order to enhance survivability when faced with adverse conditions.

9.1. Chemical toxicity

Studies of the toxicities of drugs, environmental contaminants, and other agents have commonly been carried out using acute dosing techniques in laboratory rodents. Functional and structural disorders in various systems of the treated animals have been used as biological markers to assess the potential toxicities of chemical agents. Acute toxicological studies in rodents and other species have provided an immense data base on the toxicities of thousands of drugs and environmental contaminants (e.g., Lewis and Totken, 1979).

When dosed acutely with a chemical agent such as a heavy metal, solvent, or pesticide, laboratory rodents generally show reductions in metabolic rates and become hypothermic, effects that are augmented by decreasing ambient temperatures (Figure 9.1). In the past, such reductions in body temperature

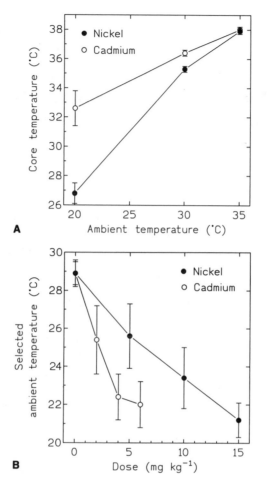

Figure 9.1. Effects of $NiCl_2$ and $CdCl_2$ on core temperature (A) and selected T_a (B) in the mouse. Doses in A: 15 mg kg^{-1} $NiCl_2$; 6 mg kg^{-1} $CdCl_2$. Data from Gordon and Stead (1986).

were viewed essentially as toxic side effects of the chemical agents. Indeed, in many toxicological studies the researchers may have blocked the hypothermia by raising the ambient temperature to assure uniformity between the control and treated groups. However, in most studies of this nature, body temperature was not measured, thus preventing any assessment of the thermoregulatory consequences of the toxic agents. Of course, not all chemical toxicants cause hypothermia. Some agents, such as 2,4-dinitrophenol, exert direct thermogenic effects by uncoupling oxidative phosphorylation and promote marked elevations in metabolic rate and body temperature (Takehiro,

Shida, and Lin, 1979). Raising the ambient temperature above thermoneu-trality can also lead to hyperthermia in rodents exposed to toxic agents. For example, rodents treated acutely with ethanol become markedly hypothermic at typical room temperatures; however, at elevated ambient temperatures (ca. 35°C), ethanol promotes faster warming rates and ensuing hyperthermic death if no intervention is made (Myers, 1981; Gordon and Mohler, 1990). Combinations of heat stress and exercise in rats treated with pyridostigmine, a cholinesterase inhibitor, lead to exhaustion and higher body temperatures compared with untreated animals (Francesconi, Hubbard, and Mager, 1984). Overall, acute toxicity leads, with few exceptions, to hypothermia in labora-tory rodents maintained at ambient temperatures below thermoneutrality.

Recently, behavioral thermoregulatory studies from my laboratory and others have found that such toxic-induced hypothermia may be an integrative thermoregulatory response, rather than simply a side effect of the chemical agent (Figure 9.1). When rodents treated with a variety of toxic chemicals are placed in a temperature gradient, they either select cooler ambient tempera-tures or fail to choose warmer temperatures to prevent the occurrence of the hypothermia (Watanabe and Suzuki, 1986; Gordon et al., 1988; Watanabe, Suzuki, and Matsuo, 1990a). This behavioral thermoregulatory effect has been found in both mice and rats treated with structurally diverse toxicants, including metals (nickel, cadmium, lead, selenite), solvents (ethanol, sulfolane), and pesticides (chlordimeform). Mice appear to be most sensitive behaviorally and select cooler temperatures during treatment. On the other hand, the responses of rats are more variable, characterized either by a re-duction in selected temperature or by no change. Yet, almost never will a treated hypothermic rodent placed in a temperature gradient select ambient temperatures above thermoneutrality, which would attenuate or block the hy-pothermia. Thus, the animal's thermoregulatory behavior seems to augment the hypothermic efficacy of the toxic agent and can be viewed as a regulated hypothermic response (see Chapter 1).

Airborne pollutants can also cause profound thermoregulatory effects in rodents. For example, exposing rats to ozone at 0.5 part per million (ppm) leads to marked reductions in body temperature and heart rate, effects that clearly are dependent on the prevailing ambient temperature (Figure 9.2). Similar to the responses to injected toxicants discussed earlier, the hypother-mic effects of ozone in rats were most pronounced at cooler temperatures and were attenuated at warmer temperatures. Interestingly, it would appear that the effects of ozone on heart rate were direct consequences of the lowering of body temperature, because bradycardia was not observed at the warmer am-bient temperature of 34°C. Thus, this represents a practical example of how

Table 9.1. *Effects of ambient temperature on lethal doses of selected chemical agents injected intraperitoneally in rats*

Compound	Lethal dose (mg kg^{-1})[a]		
	8°C	26°C	36°C
Atropine	280	420	55
Caffeine	280	280	55
Chlorpromazine	12	210	62
Digitoxin	0.7	1.0	0.5
Ethanol	1,800	1,225	800
Methanol	1,225	1,800	800
Pentobarbital	55	80	55
Toluene	530	800	225

[a]Animals were injected and then exposed to the temperatures shown for 74 hr.
Source: Data from Keplinger et al. (1959).

measurement of body temperature is crucial in the interpretation of non-thermoregulatory responses. That is, if body temperature had not been measured in this study, one might have reached the erroneous conclusion that ozone exposure directly caused bradycardia.

Why does a rodent subjected to toxic insult respond both behaviorally and autonomically to lower its body temperature? Is the response simply a total dysfunction of thermoregulatory control, or does it represent an integrated physiological response that benefits recovery from the toxic chemical? It is well known that mild hypothermia and/or relatively cool ambient temperatures in mammals and other species are quite beneficial for resisting the deleterious effects of many drugs and toxic chemicals (Doull, 1972; Harri, 1976; Selker et al., 1979; Gordon et al., 1988; Watanabe et al., 1990b). The lethal dose for many toxic chemicals, drugs, and other agents is generally reduced as ambient temperature is increased above the thermoneutral zone (Table 9.1). Although body temperature was not measured in many of these lethality studies, it is likely that raising the ambient temperature above thermoneutrality would have reduced or eliminated the toxic-induced hypothermia. Often the relationship between ambient temperature and LD$_{50}$ for a toxicant is described by an inverted V- or U-shaped curve (e.g., Fuhrman and Fuhrman, 1961; Weihe, 1973). That is, although mild hypothermia is beneficial to recovery from chemical toxicity, the thermoregulatory system of the exposed animal is nonetheless incapable of normally resisting cold stress. This is quite evident from the data in Table 9.1, where it is seen that the lethal doses for many agents were reduced when rats were exposed at an ambient temperature of 8°C. Hence, the toxicity of a chemical agent often increases during cold exposure and reaches lethality because the animal succumbs to hypothermia.

Interestingly, the effects of lower temperatures also appear to apply to natural toxins. A reduction in body temperature benefits survival following administration of some natural toxins, as well as chemical toxicants. Benton, Heckman, and Morse (1966) found that a transient lowering of body temperature in mice and guinea pigs was essential for survival following injection of bee and cobra venom. Animals kept normothermic by raising the ambient temperature experienced >90% mortality, whereas 30 min of cooling reduced mortality to 5% (Benton et al., 1966). It would also be of great interest to measure behavioral thermoregulatory parameters in rodents administered natural toxins.

How does a reduction in tissue temperature afford additional protection from chemical toxicity? It is likely that there is a variety of mechanisms in operation, but perhaps the most obvious is the Q_{10} effect of reduced temperature on toxicity. That is, the mechanisms of chemical toxicity, such as lipid peroxidation, increased membrane permeability, and conversion of a chemical to an active intermediate, are temperature-dependent phenomena. Lowering the temperature by a few degrees may provide a critical advantage in the effort to attenuate the toxicity of the chemical. Of course, hypothermia is likely to retard the detoxification and excretion of a toxicant, as has been shown with ethanol (Romm and Collins, 1987). In spite of this impairment, it appears that hypothermia is generally beneficial. Hypothermia has also been shown to protect rodents from inhaled toxicants such as ozone and formalin (Figure 9.2). In this case, the reduced body temperature and metabolic rate cause a reduction in lung ventilation, which leads to reduced intake of the airborne pollutant (Mautz and Bufalino, 1989). Hence, there appear to be two key mechanisms by which a reduced body temperature provides protection from toxic insult (Figure 9.3): attenuation in toxicity via a Q_{10} effect and reduced uptake of inhaled toxicants.

To summarize, the data indicate that rodents subjected to acute toxicity will respond behaviorally and autonomically to lower their body temperatures. Such reduction in temperature generally aids in surviving the effects of the toxic agent. This is quite an amazing biological response when we consider that the toxic chemicals that induce these effects are structurally diverse and foreign (i.e., xenobiotic) to the rodent's metabolic processes. Yet the thermoregulatory responses to toxic chemicals are similar to those seen in animals that are subjected to hypoxia and trauma and thus may be operating by similar mechanisms, as discussed later.

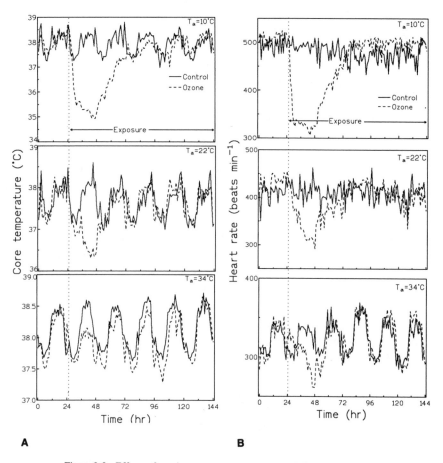

Figure 9.2. Effects of continuous exposure to ozone at 0.5 ppm on body temperature and heart rate in unanesthetized, unrestrained rats maintained at ambient temperatures of 10°C, 22°C, and 34°C. Variables measured with surgically implanted radiotelemetry transmitters. Modified from Watkinson et al. (1992).

9.1.1. Extrapolation from rodent to human

Chemical toxicity studies in rodents often are performed to give first-line estimates of potential health hazards to humans. There is an underlying assumption in many of these studies that there are sufficient similarities between species to allow extrapolation of toxicity data from experimental animals to humans. Obviously, there are many dissimilarities between the commonly used experimental animals and humans. A major endeavor in biomedical research is the development of appropriate scaling factors to facilitate interspecies comparisons (Davidson, Parker, and Beliles, 1986; Gordon, 1991).

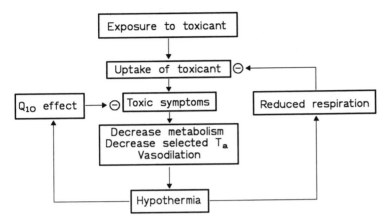

Figure 9.3. General mechanism to explain how the hypothermic response in a small rodent affords protection from acute toxic insult. Adapted from Gordon (1991) with permission from Pergamon Press.

There are marked dissimilarities in thermoregulatory responses to toxic insults in rodents and larger species that place limitations on interspecies extrapolation of toxicological data. The regulated hypothermic response commonly reported for mouse and rat is not well developed in larger mammalian species, such as humans. Although there are relatively few data, it seems clear that reductions in body temperature are limited to human subjects relative to those observed in rodents (Gordon, 1991). Only in cases of acute drug and/or alcohol overdoses combined with cold stress does one see significant hypothermia in humans comparable to the rodent responses. This finding could be attributable to a variety of autonomic and behavioral differences, but one of the most obvious factors may simply be body mass. That is, a large surface-area : body-mass ratio allows small rodents to cool relatively quickly when their metabolism is reduced, whereas a species of large size cools much slower. Thus, the protective hypothermia seen in laboratory rodents administered drugs and/or toxic chemicals is expected to be quite limited in human subjects or to take much longer to develop. It follows that this dissimilarity could lead one to underestimate the danger of acute toxicity when using data extrapolated from small rodents to large species such as humans.

9.1.2. Temperature acclimation and chemical toxicity

Most of the foregoing material deals with the effects of chemical toxicants on thermoregulation over relatively brief periods of time, generally less than

Table 9.2. *Effects of acclimation temperature on median lethal doses of various chemical agents injected intraperitoneally in mice*

	Lethal dose (mg kg^{-1})		
Compound	8°C	22°C	38°C
Benzene	101	118	115
Toluene	100	126	88
Mercuric chloride	4.8	4.5	6.5
Cadmium chloride	4.0	5.4	4.5

Source: Data from Nomiyama et al. (1980b).

2 hr. Because thermoregulation in rodents is quite sensitive to exposure to chemical toxicants, it would be reasonable to expect that various processes of temperature acclimation would also be affected. Moreover, the altered sensitivity of temperature-acclimated rodents to a chemical toxicant would be worthy of study. This latter point is especially pertinent to the study of drug and chemical toxicities. In most studies of this kind, rodents acclimated to 22°C provide the usual experimental model. Yet it is clear that human exposures can occur under a variety environmental conditions, including acclimatization to cold and warm environments. Unfortunately, relatively little work has been done in this area of study.

Cold.   As discussed earlier, cold acclimation accelerates a rodent's BMR and its ability to increase its nonshivering thermogenesis during catecholaminergic stimulation, whereas heat acclimation suppresses these responses (see Chapter 7). Considering that exposure to a chemical toxicant involves various metabolic processes (conversion of toxicant to an active metabolite, breakdown of toxicant in liver for excretion, etc.), it would seem that cold- or heat-acclimated rodents would have altered sensitivities to many chemical toxicants. For example, mice acclimated to warm and cold environments are generally more sensitive to environmental toxicants such as benzene, cadmium, and methyl parathion (Table 9.2). Cold acclimation in the rat appears to increase its sensitivity to exposure to Dioxin (Rozman and Greim, 1986). Cold acclimation also enhances the sensitivity of the mouse to methylmercury administration, whereas acclimation to 22°C and 38°C favors better survival rates (Nomiyama, Matsui, and Nomiyama, 1980b). Thus, several studies have suggested that cold acclimation exacerbates the toxicities of many chemicals. On the other hand, a 24-hr exposure to an ambient temperature of 1°C provides protection against mercuric chloride administration in rats (Burgat-Sacaze et al., 1982). In this instance, cold exposure stimulates polyurea and diuresis, which appears to protect the kidney from toxic-induced dysfunction.

The interaction between ethanol intoxication and cold acclimation is of key interest because of the frequent reports of death from hypothermia in inebriated subjects. Ethanol, like many other neurotoxicants, causes marked reductions in body temperature in laboratory rodents, an effect that is accentuated with decreasing ambient temperature (e.g., Kalant and Lê, 1984). It appears that cold acclimation can attenuate the hypothermic effects of ethanol in rodents. For example, in rats acclimated to 18°C, intraperitoneal injection of ethanol at 1.0 g kg^{-1} causes a decrease in core temperature of $\sim 1.0°C$ when they are exposed to an ambient temperature of 4°C; however, this hypothermia is almost completely blocked following 7 days of acclimation to 4°C (Lomax and Lee, 1982). Guinea pigs acclimated to 17–18°C and dosed with ethanol show better tolerance to cold exposure ($T_a = -20°C$) than do animals acclimated to 22–23°C (Huttunen and Hirvonen, 1977). It would appear that ethanol treatment in nonacclimated, cold-exposed rodents impairs their shivering thermogenesis, rendering the animals hypothermic. A cold-acclimated animal with a well-developed nonshivering thermogenic capacity is not as susceptible to hypothermia. The mechanisms of action of ethanol and other neurotoxicants on shivering and nonshivering thermogenesis remain to be determined.

Heat. Acclimation to warm temperatures may be detrimental by retarding the ability to metabolize drugs and chemical agents. Heat acclimation in rodents is also associated with reduced size of the liver and kidney, which may exacerbate the chemical toxicity (see Chapter 7). The activities of key drug-metabolizing liver enzymes, such as N-demethylase and aniline hydroxylase, decrease sharply in heat-acclimated rats (Kaplanski and Ben-Zevi, 1980). Moreover, the metabolic clearance of antipyrine, which provides a good estimate of the ability of the liver to metabolize drugs, is reduced by over 50% in heat-acclimated rats (Ben-Zevi and Kaplanski, 1980). Hexobarbitone sleep time, another index of the activity of metabolic pathways in the liver, is more than doubled following heat acclimation, suggesting impaired hepatic metabolism (Ben-Zevi and Kaplanski, 1980). It would appear that the strategy of the heat-acclimated rodent to lower its metabolic rate impacts negatively on other functions such as hepatic metabolism and renal clearance of drugs and chemical agents. This should be considered when interpreting drug metabolism studies and an animal's state of temperature acclimation.

9.2. Hypoxia and ischemia

The physiological responses to hypoxia are some of the most intensively studied topics in the neurological and cardiovascular disciplines. Tissue hypoxia during ischemia is one of the leading causes of irreversible damage to the brain and heart (e.g., stroke and heart attack, respectively). Hence, it is not surprising to find abundant studies of the effects of hypoxia and ischemia on various physiological and behavioral responses in laboratory rodents and other species.

Induction of tissue hypoxia by reducing the percentage of oxygen in the inspired air or by lowering air pressure (i.e., simulation of high altitude) results in hypothermia in rodents (Minard and Grant, 1982; Dupré, Romero, and Wood, 1988; Gordon and Fogelson, 1991b). The reduction in body temperature appears to be mediated by two principal autonomic thermoregulatory responses: (1) peripheral vasodilation, which occurs as the circulatory system attempts to deliver more blood to the oxygen-starved tissues, and (2) a reduction in metabolic rate, reflecting the reduced activity of aerobic respiration. Hypoxic rats become markedly hypothermic and hypometabolic as the ambient temperature is reduced below thermoneutrality (Figure 9.4). The lower critical temperature for elevating the metabolic rate is reduced, but the hypoxic rat can nonetheless activate facultative thermogenesis (Dupré et al., 1988). Nonshivering thermogenesis appears to be especially susceptible to hypoxia (Gautier et al., 1991), its inhibition being a major factor causing hypothermia in cold-exposed hypoxic rats.

The thermoregulatory effects of hypoxia are mediated not only systemically but also via central neural control of body temperature. For example, when rodents such as the mouse, hamster, and rat are exposed to a hypoxic atmosphere inside a temperature gradient, the selected ambient temperature is reduced, an effect that augments the hypoxia-induced hypothermia (Figure 9.5). That is, as in the chemical-toxicity response discussed earlier, hypoxia also imparts a type of regulated hypothermia. Dupré and Owen (1992) recently found that selected T_a for the rat increased from 19.6°C to 28.6°C during hypoxia, a response that was associated with a 1.6°C reduction in core temperature. The behavioral response was opposite that found in most other studies. Yet the hypoxic rat could have selected temperatures above thermoneutrality to prevent a fall in core temperature. It is interesting to note that the preference for cooler ambient temperatures during hypoxia is a common response in many endotherms and ectotherms and has been demonstrated in a variety of species, including lizard, salamander, and crayfish (Wood, Hicks, and Dupré, 1987; Dupré et al., 1988; Wood, 1991; Wood and Malvin, 1991).

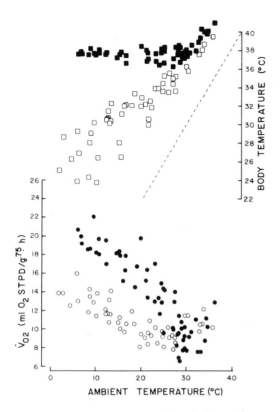

Figure 9.4. Body temperatures and metabolic rates in normoxic rats (21% O_2) and hypoxic rats (10% O_2) as functions of ambient temperature. Note reduction in lower critical ambient temperature during hypoxia. Reprinted from Dupré et al. (1988) with permission from Plenum Publishing Corp.

A favorable environment for hypoxic tissues may be created as a result of the autonomic and behavioral responses of rodents to lower their body temperatures. At the whole-animal level, survival time during acute hypoxia has been shown to be inversely related to body temperature in mice (Minard and Grant, 1982). Protective effects of hypothermia are also evident at the tissue and cellular levels. Numerous studies have found that hypoxia- and ischemia-induced tissue damage in mouse, rat, dog, and other species is attenuated as tissue temperature is reduced (Hagerdal et al., 1978; Artru and Michenfelder, 1981; Busto et al., 1987; Baldwin et al., 1991). For example, the neuronal damage to the brain in hypoxic rats is clearly attenuated by lowering the brain temperature during and/or immediately after ischemia (Busto et al., 1987). White, Feuerstein, and Barone (1992) recently reported that ischemic damage

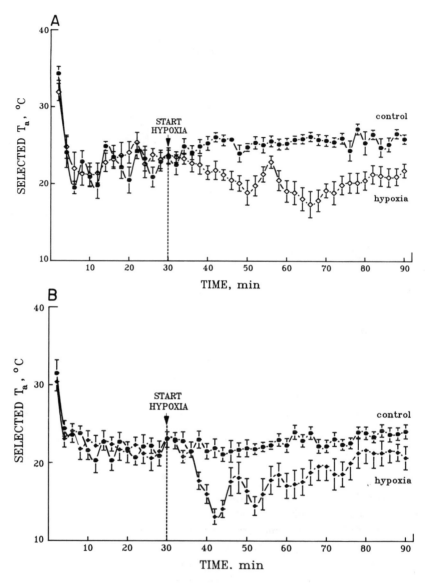

Figure 9.5. Behavioral thermoregulatory responses to hypoxia in hamsters (A) and rats (B). Rats and hamsters were exposed to 7.3% and 6.7% oxygen, respectively, for 60 min while maintained in a temperature gradient. Body temperatures at end of exposure period were 38.2°C and 35.4°C for normoxic and hypoxic rats, and 37.4°C and 33.5°C for normoxic and hypoxic hamsters. Reprinted from Gordon and Fogelson (1991b) with permission from The American Physiological Society.

to the rat brain following transient occlusion of the middle cerebral artery could be completely prevented by lowering the brain temperature by 8°C. Hence, lowering the body temperature reduces tissue oxygen requirements, thus prolonging survival and affording protection to the hypoxic tissues.

There are interesting interactions between the thermoregulatory reflexes of rodents subjected to hypoxia and the efficacies of antiischemic drugs and related treatments. Because of its unique circulatory structure, the Mongolian gerbil has become a popular species in which to assess the potential benefits of drugs intended as prophylaxis for stroke. This species lacks a circle of Willis, and so blood is delivered to its brain solely by the internal carotid arteries. Thus, complete global brain ischemia can be brought on quickly by experimental occlusion of the common carotid arteries. Generally, an antiischemic drug is given to the gerbil before, during, and/or after clamping the carotids for a relatively brief period, such that CNS damage is evoked, without killing the animal.

Several studies have reported that *N*-methyl-*D*-aspartate-channel antagonists such as MK-801 attenuate CNS damage in the ischemic gerbil brain. The prophylactic effect of MK-801 is especially prominent in the vulnerable CA1 pyramidal neurons of the hippocampus (Busto et al., 1987). Buchan and Pulsinelli (1990) recognized that MK-801 treatment also caused profound hypothermia, a response that could affect the protective action of the drug. Indeed, they found that if the gerbil was kept normothermic for 8 hr after ischemia and MK-801 administration, the brain damage was as severe as that found in ischemic gerbils not given MK-801. Moreover, if after ischemia the body temperature was lowered to the same level as that during the MK-801 treatment, the protection to the CNS was similar to that in MK-801-treated animals. This clearly illustrates a problem in using thermally labile rodents such as the gerbil in these types of drug studies. It will also be recalled that an equivalent drug-induced hypothermia in the gerbil is unlikely to occur in an adult human subject, which further compounds the danger in trying to extrapolate from such drug studies in laboratory rodents to adult humans, although their relevance to human infants is perhaps more likely (see Section 9.1.1).

9.3. Trauma and shock

Rodents are often used as experimental models in the study of trauma, which is essentially defined as the sum of the physiological responses to a physical injury. H. B. Stoner's studies of rats were instrumental in furthering our understanding of the effects of trauma on thermoregulatory processes (Stoner,

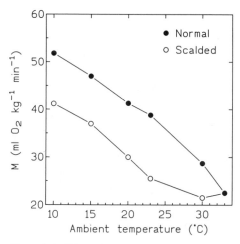

Figure 9.6. Effect of burn injury (20% of skin scalded for 30 sec at 83°C) on ambient-temperature–metabolic-rate profile of the rat. Metabolic rate measured 4.0–5.5 hr after injury. Data from Stoner (1969).

1968, 1971). As in the case of their response to chemical toxicity discussed earlier, rats undergo reductions in metabolic rate and body temperature when subjected to various types of physical trauma, such as skin burns and transient bilateral hindlimb ischemia. The threshold ambient and hypothalamic temperatures for activation of shivering are markedly reduced in traumatized rats (Stoner, 1971, 1972). The lower critical temperature for elevating the metabolic rate is also reduced following a burn injury; however, facultative thermogenesis is nonetheless operative (Figure 9.6). That is, a traumatized rat appears capable of elevating its metabolic rate, but does not do so until the ambient temperature is 7°C cooler (relative to the response of a control rat). These data suggest that as in the case of responses to chemical toxicity and hypoxia, the set-point for control of body temperature is reduced by trauma. A rat's body temperature generally returns to baseline by 24 hr after the trauma (Stoner, 1968), but its ability to thermoregulate in acute cold exposure is impaired for at least 48 hr (Stoner, 1968).

Is the hypothermic response in traumatized rodents beneficial? The metabolic alterations during the onset of and recovery from trauma can be quite complex, yet it appears that cooling to a certain point is beneficial. Stoner (1968) and others have concluded that there is an optimal ambient temperature for survival after skin burning, usually 20°C. This ambient temperature is at a level that "allows the body temperature to fall at an optimum rate to a level which is not so low that unaided recovery is impossible yet one which confers advantages by limiting the injury by reducing the metabolic demands of the tissues.'' In mice, recovery from burns can be affected by a variety of factors, such as food availability, the bacterial environment, and the presence of fur (Markley et al., 1973). In normally furred mice, survival is enhanced by lowering the ambient temperature from 31°C to 25°C, whereas survival among furless animals is greatly improved by elevating the ambient temperature.

The interactions between thermoregulatory responses to trauma and their possible therapeutic benefits have been demonstrated in the neural, cardiovascular, pancreatic, and renal systems. For example, physical damage to the rat brain is partially alleviated by reducing the brain temperature through administration of an anesthetic (Clifton et al., 1991). Rats subjects to hemorrhagic shock via acute blood loss normally become hypothermic and display severe ECG arrhythmias and loss of blood pressure (Tanaka et al., 1983). However, survival time following the blood loss is prolonged significantly by maintaining the animals in a hypothermic state. During insulin shock, the sudden lowering of blood glucose results in a marked reduction in body temperature in the rat and other mammals (Buchanan et al., 1991). Again, as has been found with stress due to other forms of trauma, survival time during insulin shock is reduced if the animals are kept normothermic, whereas animals allowed to become hypothermic experience 100% survival (Figure 9.7). Indeed, normothermic rats survived no longer than 4 hr after plasma glucose was decreased below 40 mg dl^{-1}, whereas hypothermic rats survived similar blood glucose levels for more than 8 hr (Buchanan et al., 1991). During renal failure, the development of uremia and the accumulation of toxic metabolites in the blood leads to an array of pathological events, including hypothermia. Acute uremia is associated with a reduced metabolic rate and hypothermia in rats (Om and Hohenegger, 1980). Induction of uremia by injection of urine into a mouse leads to a reduction in body temperature and a preference for cooler ambient temperatures, suggesting a regulated hypothermic response (Gordon, 1990b). Given the aforementioned discussion, it would seem that the hypothermia accompanying uremia would attenuate the pathological effects of uremia in the CNS and other organs.

9.4. Hypergravity

Understanding the physiological responses of humans subjected to hypergravic fields has always been a critical aspect of aerospace research. Such studies in rodents are generally carried out in a centrifuge, in which the animal is protected from increased convective heat loss by appropriate

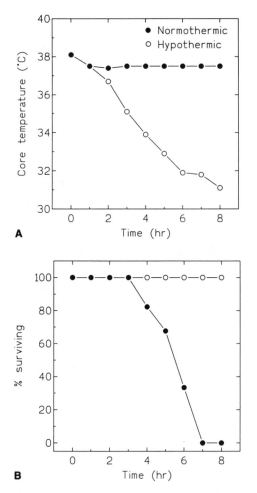

Figure 9.7. Beneficial effects of hypothermia on survival following insulin shock in the rat. Plasma glucose was reduced to 30–40 mg dl^{-1} by continuous insulin infusion beginning after time zero. Data from Buchanan et al. (1991).

shielding. When the gravitational field is increased from 1 *g* to ≥1.5 *g*, increased pressure on the skin and internal tissues appears to be a main cause of the altered responses of the cardiovascular, respiratory, and thermoregulatory systems in rodents.

Changes in core and skin temperatures during acceleration have been measured in the rat and other species (Oyama, Platt, and Holland, 1971; Fuller, Horowitz, and Horwitz, 1977; Giacchino, Horwitz, and Horowitz, 1979; Ohara et al., 1982). At relatively cool ambient temperatures (i.e., below the

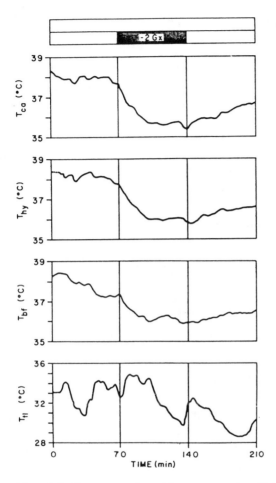

Figure 9.8. Time courses of carotid artery temperature (T_{ca}), hypothalamic temperature (T_{hy}), brown-fat temperature (T_{bf}), and tail skin temperature (T_{tl}) in a loosely restrained rat subjected to acceleration equivalent to an increase in the gravitational field from 1 to 2 G for 70 min at night. Hypergravic exposure began at 70 and ended at 140 min. Reprinted from Fuller et al. (1991) with permission from The American Physiological Society.

thermoneutral zone) there is commonly a reduction in body temperature concomitant with an elevation in tail skin temperature during hypergravic exposure (Figure 9.8). This would imply a regulated reduction in body temperature, because the rat apparently is vasodilating to increase heat loss. However, this supposition cannot be verified without measurements of behavioral thermoregulatory responses, which apparently have not been made in rodents subjected to hypergravic fields. The greatest reduction in core temperature in the rat during acceleration occurs at night, when temperature is normally elevated (Fuller, Griffin, and Horowitz, 1991). Thus, the circadian thermoregulatory rhythm is a critical factor in the hypothermic response to a hypergravic environment.

Interestingly, the reduction in body temperature may well be a beneficial response, like that to the other traumatic stimuli discussed earlier. Ohara et al. (1982) measured body and tail skin temperatures in rats subjected to severe hypergravic fields of up to 5.8 g over an ambient temperature range of 15–35°C. Rats subjected to those fields at 15°C exhibited peripheral vasodilation and reductions in core temperatures. Those thermoregulatory responses were associated with 100% survival in the hypergravic environment. On the other hand, rats exposed at the warmer temperatures showed vasoconstriction of the tail and increases in core temperatures; all those animals died within 20 min of the start of the hypergravic exposure. The mechanism behind the rise in core temperature is not clear. That is, if hypothermia is beneficial, why is there vasoconstriction rather than vasodilation in the tail? Rats subjected to hypergravic fields at high ambient temperatures apparently are unable to salivate (Ohara et al., 1982). But even if salivation were possible, the restrained animals would be unable to groom saliva to increase their evaporative water loss (see Chapter 4). Hypothermia has also been found to protect mice from the lethal effects of hypergravitational fields (Felice and Miller, 1972). It would seem, then, that a lower body temperature is beneficial for combating the deleterious effects of hypergravity in rodents. It is possible that the circulatory disturbances and restrictions in oxygen delivery caused by acceleration are ameliorated by lowering the body temperature (Ohara et al., 1982). Whereas the initial exposure to hypergravity reduces rodents' core temperatures, it should be noted that chronic exposure to 2.0 g for several weeks leads to elevations in metabolic rates in hamsters, rats, and guinea pigs (Pace and Smith, 1983).

9.5. Regulated hypothermia: a generalized protective mechanism?

Clearly, many adverse stimuli will generally produce a reduction in body temperature in rodents, particularly rats and mice. Incredibly, the hypothermic response is evoked by a variety of stimuli, including chemical toxicants, hypoxia, ischemia, physical trauma, hypergravity, and others. The hypothermic response generally improves the animal's ability to survive or resist the environmental insult. Moreover, in many instances in which behavioral thermoregulation has been monitored, a reduction in selected ambient temperature occurs simultaneously with the reduced body temperature, which suggests that the hypothermia is a regulated thermoregulatory response, not simply a dysfunction of homeostatic processes.

Just as an infected animal selects warmer ambient and body temperatures during fever, a response that appears to improve its ability resist infectious organisms (see Chapter 2), it seems that a traumatized or otherwise injured rodent prefers lower body temperatures, a response that also improves its recovery. Is this hypothermic response a specific adaptation of the thermoregulatory system or simply a breakdown or failure of behavioral and autonomic control? Considering that the thermoregulatory system is usually very stable, it is unusual to find a behaviorally and autonomically mediated response to lower the body temperature. Interestingly, there is support for the concept of evolutionary development of a generalized hypothermic response. S. C. Wood and colleagues have postulated that regulated hypothermia in response to stress (toxins and pH, osmotic, oxygen, and other stresses) is basically a general protective response present in both lower and upper vertebrates that could be mediated by generalized neurohumoral agents such as vasopressin (Wood and Malvin, 1991).

Temperature regulation in homeotherms has evolved to maintain stable internal temperatures, thus providing the physiological and behavioral processes for an optimal thermal environment over a relatively wide range of environmental conditions. Yet, for a rodent subjected to extremely adverse conditions (chemical toxicity, hypoxia, trauma, etc.), the normally stable temperature of 37°C can become a liability to its survival. It appears that a hypothermic strategy comes into play as a means of enhancing survival and reducing tissue damage. Researchers not trained in thermal physiology should nevertheless be cognizant of the hypothermic response. Its activation clearly can impact on the functioning of many physiological systems. Moreover, if body temperature is not measured in stress-related studies, researchers may reach erroneous conclusions regarding the effect, or lack of effect, of the stressor under study.

References

Abelenda, M., and M. L. Puerta (1987). Inhibition of diet-induced thermogenesis during pregnancy in the rat. *Pflügers Arch.*, 409:314–17.

Abrams, R., and H. T. Hammel (1964). Hypothalamic temperature in unanesthetized rats during feeding and sleeping. *Am. J. Physiol.*, 206:641–6.

—— (1965). Cyclic variations in hypothalamic temperature in unanesthetized rats. *Am. J. Physiol.*, 208:698–702.

Abrams, R., J. A. J. Stolwijk, H. T. Hammel, and H. Graichen (1965). Brain temperature and brain blood flow in unanesthetized rats. *Life Sci.*, 4:2399–410.

Ackerman, D., and T. A. Rudy (1980). Thermoregulatory characteristics of neurogenic hyperthermia in the rat. *J. Physiol.*, 307:59–70.

Adachi, A., M. Funahashi, and J. Ohga (1991). Hepatic thermogenesis in relation to food intake in the conscious rat. *Brain Res. Bull.*, 27:529–33.

Adamsons, K., E. Blumberg, and I. Joelsson (1969). The effect of ambient temperature upon post-natal changes in oxygen consumption of the guinea-pig. *J. Physiol.*, 202:261–9.

Adels, L. E., and M. Leon (1986). Thermal control of mother–young contact in Norway rats: factors mediating the chronic elevation of maternal temperature. *Physiol. Behav.*, 36:183–96.

Adler, M. W., E. B. Geller, C. E. Rosow, and J. Cochin (1988). The opioid system and temperature regulation. *Annu. Rev. Pharmacol.*, 28:429–49.

Adolph, E. F. (1947). Tolerance to heat and dehydration in several species of mammals. *Am. J. Physiol.*, 151:564–75.

—— (1957). Ontogeny of physiological regulations in the rat. *Q. Rev. Biol.*, 32:89–137.

Adolph, E. F., and J. W. Larow (1951). Acclimatization to cold air; hypothermia and heat production in the golden hamster. *Am. J. Physiol.*, 166:62–74.

Adolph, E. F., and J. Richmond (1955). Rewarming from natural hibernation and from artificial cooling. *J. Appl. Physiol.*, 8:48–58.

Akins, C., and D. D. Thiessen (1990). Lipopolysaccharide decreased ambient temperature preference in Mongolian gerbils, *Meriones unguiculatus*. *Percept. Mot. Skills*, 71:1177–8.

Akins, C., D. Thiessen, and R. Cocke (1991). Lipopolysaccharide increases ambient temperature preference in C57BL/6J adult mice. *Physiol. Behav.*, 50:461–3.

225

Alberts, J. R. (1978). Huddling by rat pups: group behavioral mechanisms of temperature regulation and energy conservation. *J. Comp. Physiol. Psychol.,* 92:231–45.

Alfoldi, P., G. Rubicsek, G. Cserni, and F. Obál, Jr. (1990). Brain and core temperatures and peripheral vasomotion during sleep and wakefulness at various ambient temperatures in the rat. *Pflügers Arch.,* 417:336–41.

Al-Hilli, F., and E. A. Wright (1988a). The effects of environmental temperature on the growth of hair in the mouse. *J. Therm. Biol.,* 13:21–4.

(1988b). The effects of environmental temperature on the body temperature and ear morphology of the mouse. *J. Therm. Biol.,* 13:197–9.

Altman, P. L., and D. S. Dittmer (eds.) (1962). *Growth: Biological Handbook.* Federation of American Societies of Experimental Biology, Washington, DC.

Anantharaman-Barr, H. G., and J. Decombaz (1989). The effect of wheel running and the estrous cycle on energy expenditure in female rats. *Physiol. Behav.,* 46:259–63.

Anderson, C. R., and E. M. McLachlan (1991). The time course of the development of the sympathetic innervation of the vasculature of the rat tail. *J. Auton. Nerv. Syst.,* 35:117–32.

Anderson, G. L., W. A. Volkert, and X. J. Musacchia (1971). O_2 consumption, electrocardiogram, and spontaneous respiration of hypothermic hamsters. *Am. J. Physiol.,* 221:1774–8.

Andjus, R. K., and J. E. Lovelock (1955). Reanimation of rats from body temperatures between O and 1°C by microwave diathermy. *J. Physiol.,* 128:541–6.

Andrews, J. F., D. Richard, G. Jennings, and P. Trayhurn (1986). Brown adipose tissue thermogenesis during pregnancy in mice. *Ann. Nutr. Metab.,* 30:87–93.

Andrews, J. F., and E. M. Ryan (1976). Oxygen consumption related to ambient temperature in rats of different thyroid status. *J. Physiol.,* 263:258–9.

Arieli, A., and A. Chinet (1986). Thyroid status and noradrenaline-induced regulatory thermogenesis in heat acclimated rats. *Horm. Metab. Res.,* 18:103–6.

Arjamaa, O., and K. Y. Lagerspetz (1979). Postnatal development of shivering in the mouse. *J. Therm. Biol.,* 4:35–9.

Armitage, G., R. B. Harris, G. R. Hervey, and G. Tobin (1984). The relationship between energy expenditure and environmental temperature in congenitally obese and nonobese Zucker rats. *J. Physiol.,* 350:197–207.

Arnold, J., J. LeBlanc, G. Cote, H. Lalonde, and D. Richard (1986). Exercise suppression of thermoregulatory thermogenesis in warm- and cold-acclimated rats. *Can. J. Physiol. Pharmacol.,* 64:922–6.

Artru, A. A., and J. D. Michenfelder (1981). Influence of hypothermia or hyperthermia alone or in combination with pentobarbital or phenytoin on survival time in hypoxic mice. *Anesth. Analg.,* 60:867–70.

Aschoff, J. (1981). Thermal conductance in mammals and birds: its dependence on body and size and circadian phase. *Comp. Biochem. Physiol.,* 69A:611–19.

Attah, M. Y., and E. L. Besch (1977). Estrous cycle variations of food and water intake in rats in the heat. *J. Appl. Physiol.,* 42:874–7.

Avery, D. D. (1972). Thermoregulatory effects of intrahypothalamic injections of adrenergic and cholinergic substances at different environmental temperatures. *J. Physiol.,* 220:257–66.

Baciu, I., A. Olteanu, T. Prodan, M. Baiescu, and A. Vaida (1985). Changes of phagocytic biological rhythm by reduction of circadian times and by influences upon hypothalamus. In *Neuroimmunomodulation* (pp. 142–9), Proceedings, 1st International Workshop, Bethesda, MD.

Baker, M. A. (1982). Brain cooling in endotherms in heat and exercise. *Annu. Rev. Physiol.*, 44:85–96.

Bakken, G. S. (1991). Time-resolved respirometry: equations for the simultaneous measurement of all respiratory gases and the calibration of oxygen consumption using variable inert gas flow rates. *J. Therm. Biol.*, 16:313–15.

Baldwin, B. A. (1968). Behavioural thermoregulation in mice. *Physiol. Behav.* 3:401–7.

Baldwin, W. A., J. R. Kirsch, P. D. Hurn, W. S. P. Toung, and R. J. Traystman (1991). Hypothermic cerebral reperfusion and recovery from ischemia. *Am. J. Physiol.*, 261:H774–81.

Balmagiya, T., and S. J. Rozovski (1983). Age-related changes in thermoregulation in male albino rats. *Exp. Gerontol.*, 18:199–210.

Banet, M. (1988). Long-term cold adaptation in the rat. *Comp. Biochem. Physiol.*, 89A:137–40.

Banet, M., and H. Hensel (1976). The interaction between cutaneous and spinal thermal inputs in the control of oxygen consumption in the rat. *J. Physiol.*, 260:461–73.

Banet, M., H. Hensel, and H. Liebermann (1978). The central control of shivering and non-shivering thermogenesis in the rat. *J. Physiol.*, 283:569–84.

Banet, M., and J. J. Seguin (1970). Effects of preoptic cooling in rats acclimated to 21 and 4°C. *J. Appl. Physiol.*, 29:385–8.

Barnett, S. A. (1959). The skin and hair of mice living at a low environmental temperature. *Q. J. Exp. Physiol.*, 44:35–42.

(1973). Maternal processes in the cold-adaptation of mice. *Biol. Rev.*, 48:477–508.

Barnett, S. A., and E. M. Widdowson (1965). Organ-weights and body-composition in mice bred for many generations at −3°C. *Proc. R. Soc. Lond. [Biol.]*, 162:502–15.

Barney, C. C., J. S. Williams, and D. H. Kuiper (1991). Thermal dehydration-induced thirst in rats: role of angiotensin II. *Am. J. Physiol.*, 261:R1171–5.

Bartlett, R. G., Jr. (1959). Effects of restraint on oxygen consumption of the cold exposed guinea pig. *J. Appl. Physiol.*, 14:46–8.

Bartunkova, R., L. Janský, and J. Mejsnar (1971). Nonshivering thermogenesis and cold adaptation. In L. Janský (ed.), *Nonshivering Thermogenesis* (pp. 39–56). Prague: Academia.

Batchelder, P., R. O. Kinney, L. Demlow, and C. B. Lynch (1983). Effects of temperature and social interactions on huddling behavior in *Mus musculus*. *Physiol. Behav.*, 31:97–102.

Beckman, A. L., and H. J. Carlisle (1969). Effect of intrahypothalamic infusion of acetylcholine on behavioural and physiological thermoregulation in the rat. *Nature*, 221:561–2.

Bedrak, E., K. Fried, V. Samoiloff, and U. A. Sod-Moriah (1977). The effect of acclimation to elevated ambient temperature on progesterone levels in blood and corpora lutea of cycling and four-day pregnant rats. *J. Therm. Biol.*, 2:83–7.

Bellvé, A. R. (1973). Development of mouse embryos with abnormalities induced by parental heat stress. *J. Reprod. Fertil.*, 35:393–403.

Benedict, F. G., and G. MacLeod (1929). The heat production of the albino rat. II. Influence of environmental temperature, age, and sex; comparison with the basal metabolism of man. *J. Nutr.*, 1:367–98.

Benjanian, M., D. A. Finn, P. J. Syapin, and R. L. Alkana (1987). Rectal and brain temperatures in ethanol intoxicated mice. *Psychopharmacology*, 92:301–7.

Benton, A. W., R. A. Heckman, and R. A. Morse (1966). Environmental effects on venom toxicity in rodents. *J. Appl. Physiol.*, 21:1228–30.

Ben-Zevi, Z., and J. Kaplanski (1980). Effects of chronic heat exposure on drug metabolism in the rat. *J. Pharm. Pharmacol.*, 32:368–9.

Benzinger, T. H., and C. Kitzinger (1963). Gradient layer calorimetry and human calorimetry. In J. D. Hardy (ed.), *Temperature, Its Measurement and Control in Science and Industry*, Vol. III, Pt. 3 (pp. 87–109). New York: Reinhold.

Berman, E., D. House, and H. B. Carter (1990). Effect of ambient temperature and running wheel activity on the outcome of pregnancy in CD-1 mice. *Terat. Carcin. Mutagen.*, 10:11–20.

Blackshaw, A. W. (1977). Temperature and seasonal influences. In A. S. Johnson and W. R. Gomez (eds.), *Testes*, Vol. 4 (pp. 517–46). New York: Academic Press.

Blatteis, C. M. (1974). Influence of body weight and temperature on the pyrogenic effects of endotoxin in guinea pig. *Toxicol. Appl. Pharmacol.*, 29:249–58.

 (1976). Fever: exchange of shivering by nonshivering pyrogenesis in cold-acclimated guinea pigs. *J. Appl. Physiol.*, 40:29–34.

Blatteis, C. M., and M. Banet (1986). Autonomic thermoregulation after separation of the preoptic area from the hypothalamus in rats. *Pflügers Arch.*, 406:480–4.

Blatteis, C. M., W. S. Hunter, J. M. Wright, R. A. Ahokas, J. Llanos-Q., and T. A. Mashburn, Jr. (1987). Thermoregulatory responses of guinea pigs with anteroventral third ventricle lesions. *Can. J. Physiol. Pharmacol.*, 65:1261–6.

Blatteis, C. M., and K. A. Smith (1980). Behavioral fever induced in guinea-pigs by intrapreoptic pyrogen. *Experientia*, 36:1086–8.

Blaxter, K. L. (1989). *Energy Metabolism in Animals and Man*. Cambridge University Press.

Bligh, J., and K. G. Johnson (1973). Glossary of terms for thermal physiology. *J. Appl. Physiol.*, 35:941–61.

Blumberg, M. S., J. A. Mennella, and H. Moltz (1987). Hypothalamic temperature and deep body temperature during copulation in the male rat. *Physiol. Behav.*, 39:367–70.

Boulant, J. A. (1991). Thermoregulation. In P. Mackowiak (ed.), *Fever: Basic Mechanisms and Management* (pp. 1–22). New York: Raven Press.

Boulant, J. A., M. C. Curras, and J. B. Dean (1989). Neurophysiological aspects of thermoregulation. In L. C. H. Wang (ed.), *Comparative and Environmental Physiology. Vol. 4: Animal Adaptation to Cold* (pp. 117–60). London: Springer-Verlag.

Boulant, J. A., and J. B. Dean (1986). Temperature receptors in the central nervous system. *Annu. Rev. Physiol.*, 48:639–54.

Boulant, J. A., and N. L. Silva (1989). Multisensory hypothalamic neurons may explain interactions among regulatory systems. *News Physiol. Sci.*, 4:245–8.

Bradley, S. R., and D. R. Deavers (1980). A re-examination of the relationship between thermal conductance and body weight in mammals. *Comp. Biochem. Physiol.*, 65A:465–76.

Bramante, P. O. (1968). Energy metabolism of the albino rat in minimal levels of spontaneous muscular activity. *J. Appl. Physiol.*, 24:11–16.

Briese, E. (1985). Rats prefer ambient temperatures out of phase with their body temperature circadian rhythm. *Brain Res.*, 345:389–93.

Briese, E., and M. Cabanac (1991). Stress hyperthermia: physiological arguments that it is a fever. *Physiol. Behav.*, 49:1153–7.

Brittain, R. T., and S. L. Handley (1967). Temperature changes produced by the injection of catecholamines and 5-hydroxytryptamine into the cerebral ventricles of the conscious mouse. *J. Physiol.*, 192:805–13.

Brooks, D. E. (1973). Epididymal and testicular temperature in the unrestrained conscious rat. *J. Reprod. Fertil.*, 35:157–60.

Brown, D., G. Livesey, and M. J. Dauncey (1991). Influence of mild cold on the components of 24-hour thermogenesis in rats. *J. Physiol.*, 441:137–54.

Brown, R. T., and J. G. Baust (1980). Time course of peripheral heterothermy in a homeotherm. *Am. J. Physiol.*, 239:R126–9.

Brown, S. J., C. V. Gisolfi, and F. Mora (1982). Temperature regulation and dopaminergic systems in the brain: does the substantia nigra play a role? *Brain Res.*, 234:275–86.

Brück, K., and P. Hinckel (1980). Thermoregulatory noradrenergic and serotonergic pathways to hypothalamic units. *J. Physiol.*, 304:193–202.

(1990). Thermoafferent networks and their adaptive modifications. In E. Schönbaum and P. Lomax (eds.), *Thermoregulation: Physiology and Biochemistry* (pp. 129–52). New York: Pergamon Press.

Brück, K., and B. Wünnenberg (1966). Influence of ambient temperature in the process of replacement of nonshivering by shivering thermogenesis during postnatal development. *Fed. Proc.*, 25:1332–6.

Brück, K., W. Wünnenberg, H. Gallmier, and B. Ziehm (1970). Shift of threshold temperature for shivering and heat polypnea as a mode of thermal adaptation. *Pflügers Arch.*, 321:159–72.

Brück, K., and E. Zeisberger (1978). Significance and possible central mechanisms of thermoregulatory threshold deviations in thermal adaptation. In L. C. H. Wang and J. W. Hudson (eds.), *Strategies in Cold, Natural Torpidity and Thermogenesis* (pp. 655–94). New York: Academic Press.

Buchan, A., and W. A. Pulsinelli (1990). Hypothermia but not the N-methyl-D-aspartate antagonist, MK-801, attenuates neuronal damage in gerbils subjected to transient global ischemia. *J. Neurosci.*, 10:311–16.

Buchanan, T. A., P. Cane, C. C. Eng, G. F. Sipos, and C. Lee (1991). Hypothermia is critical for survival during prolonged insulin-induced hypoglycemia in rats. *Metabolism*, 40:330–4.

Buhler, H. U., M. Da Prado, W. Haefely, and G. B. Picotti (1978). Plasma adrenaline, noradrenaline and dopamine in man and different animal species. *J. Physiol.*, 276:311–20.

Burgat-Sacaze, V., J. Braun, P. Benard, B. Eghbali, and A. Rico (1982). Protection against mercuric chloride nephrotoxicity by cold exposure in rats. *Toxicology Lett.*, 10:151–6.

Busto, R., W. D. Dietrich, M. Globus, I. Valdes, P. Scheinberg, and M. D. Ginberg (1987). Small differences in intraischemic brain temperature critically determine the extent of ischemic neuronal injury. *J. Cereb. Blood Flow Metab.*, 7:729–38.

Cabanac, A., and E. Briese (1991). Handling elevates the colonic temperature of mice. *Physiol. Behav.*, 51:95–8.

Cabanac, M. (1975). Temperature regulation. *Annu. Rev. Physiol.*, 38:415–39.

(1986). Keeping a cool head. *News Physiol. Sci.*, 1:41–4.

Calder, W. A. (1981). Scaling of physiological processes in homeothermic animals. *Annu. Rev. Physiol.*, 43:301–22.

Caldwell, F. T., H. T. Hammel, and F. A. Dolan (1966). A calorimeter for simultaneous determination of heat production and heat loss in the rat. *J. Appl. Physiol.*, 21:1665–71.

Campbell, B. A., and G. S. Lynch (1967). Activity and thermoregulation during food deprivation in the rat. *Physiol. Behav.*, 2:311–13.

(1968). Influence of hunger and thirst on the relationship between spontaneous activity and body temperature. *J. Comp. Physiol. Psychol.*, 65:492–8.

Cantor, A., and E. Satinoff (1976). Thermoregulatory responses to intraventricular norepinephrine in normal and hypothalamic-damaged rats. *Brain Res.*, 108:125–41.

Caputa, M., W. Kadzie, and J. Narebski (1983). Cerebral temperature regulation in resting and running guinea-pigs (*Cavia porcellus*). *J. Therm. Biol.*, 8:265–72.

Caputa, M. and A. Kamari (1991). Effect of warm rearing on temperature regulation in resting and exercising rats. *J. Therm. Biol.*, 16:357–61.

Caputa, M., A. Kamari, and M. Wachulec (1991). Selective brain cooling in rats resting in heat and during exercise. *J. Therm. Biol.*, 16:19–24.

Caputa, M., E. Wasilewska, and E. Swiecka (1985). Hyperthermia and exercise performance in guinea-pigs (*Cavia porcellus*). *J. Therm. Biol.*, 10:217–20.

Carlisle, H. J. (1968). Peripheral thermal stimulation and thermoregulatory behavior. *J. Comp. Physiol. Psychol.*, 66:507–10.

Carlisle, H. J., and P. U. Dubuc (1984). Temperature preference of genetically obese (ob/ob) mice. *Physiol. Behav.*, 33:899–902.

Carlisle, H. J., C. W. Wilkinson, M. L. Laudenslager, and L. D. Keith (1979). Diurnal variation of heat intake in ovariectomized, steroid-treated rats. *Horm. Behav.*, 12:232–42.

Cassuto, Y. (1968). Metabolic adaptations to chronic heat exposure in the golden hamster. *Am. J. Physiol.*, 214:1147–51.

(1970). Metabolic responses of heat-acclimated hamsters during deacclimation. *Am. J. Physiol.*, 218:1560–2.

Cassuto, Y., R. Chayoth, and T. Rabi (1970). Thyroid hormone in heat-acclimated hamsters. *Am. J. Physiol.*, 218:1287–90.

Castella, J., and M. Alemany (1985). Sex differences in the thermogenic response of the rat to a "cafeteria" diet. *IRCS Med. Sci.*, 13:586–7.

Chaffee, R. R. J., and J. C. Roberts (1971). Temperature acclimation in birds and mammals. *Annu. Rev. Physiol.*, 33:155–202.

Chap, Z., and E. Bedrak (1983). Interrelationship between pituitary-testicular axis activity and raised environmental temperature in the rat. *J. Endocrinol.*, 97:193–200.

Chatfield, P. O., A. F. Battista, C. P. Lyman, and J. P. Garcia (1948). Effects of cooling on nerve conduction in a hibernator (golden hamster) and non-hibernator (albino rat). *Am. J. Physiol.*, 155:179–85.

Chaudhry, A. P., F. Halberg, C. E. Keenan, R. N. Harner, and J. J. Bittner (1958). Daily rhythms in rectal temperature and in epithelial mitoses of hamster pinna and pouch. *J. Appl. Physiol.*, 12:221–4.

Chayoth, R., D. Kleinman, J. Kaplanski, and U. A. Sod-Moriah (1984a). Renal clearance of urea, inulin, and *p*-aminohippurate in heat-acclimated rats. *J. Appl. Physiol.*, 57:731–2.

Chayoth, R., A. F. Nakhooda, P. Poussier, and E. B. Marliss (1984b). Glucoregulatory and metabolic responses to heat exposure in rats. *Am. J. Physiol.*, 246:E465–70.

Chevillard, L., R. Bertin, and M. Cadot (1967). The influence of rearing temperature on some physiological characteristics of small laboratory animals (homeotherms). In M. L. Conalty (ed.), *Husbandry of Laboratory Animals* (pp. 395–445). London: Academic Press.

Chevillard, L., R. Porter, and M. Cadot (1963). Growth rate of rats born at 5 and 30°C. *Fed. Proc.*, 22:699–703.

Chew, R. M. (1951). The water exchanges of some small mammals. *Ecol. Monogr.*, 21:215–25.

Chiu, C. C., and A. C. L. Hsieh (1960). A comparative study of four means of expressing the metabolic rate of rats. *J. Physiol.*, 150:694–706.

Christman, J. V., and C. V. Gisolfi (1980). Effects of repeated heat exposure on hypothalamic sensitivity to norepinephrine. *J. Appl. Physiol.*, 49:942–5.

(1985). Heat acclimation: role of norepinephrine in the anterior hypothalamus. *J. Appl. Physiol.*, 58:1923–8.

Clark, R. V. (1971). Behavioral thermoregulation by the white rat at high ambient temperatures. *J. Exp. Zool.*, 178:387–92.

Clark, W. G. (1979). Changes in body temperature after administration of amino acids, peptides, dopamine, neuroleptics and related agents. *Neurosci. Biobehav. Rev.*, 3:179–231.

(1991). Antipyretics. In P. Mackowiak (ed.), *Fever: Basic Mechanisms and Management* (pp. 297–340). New York: Raven Press.

Clark, W. G., and Y. L. Clark (1980). Changes in body temperature after administration of acetylcholine, histamine, morphine, prostaglandins and related agents. *Neurosci. Biobehav. Rev.*, 4:175–240.

(1981). Changes in body temperature after administration of antipyretics, LSD, Δ^9-THC, CNS depressants and stimulants, hormones, inorganic ions, gases, 2,4-DNP and miscellaneous agents. *Neurosci. Biobehav. Rev.*, 5:1–136.

Clark, W. G., and J. M. Lipton (1983). Brain and pituitary peptides in thermoregulation. *Pharmacol. Ther.*, 22:249–97.

(1985). Changes in body temperature after administration of amino acids, peptides, dopamine, neuroleptics and related agents: II. *Neurosci. Biobehav. Rev.*, 9:299–371.

(1986). Changes in body temperature after administration of adrenergic and serotonergic agents and related drugs including antidepressants: II. *Neurosci. Biobehav. Rev.*, 10:153–220.

Clarkson, D. P., C. L. Schatte, and J. P. Jordan (1972). Thermal neutral temperature of rats in helium-oxygen, argon-oxygen, and air. *Am. J. Physiol.,* 222: 1494–8.

Clifton, G. L., J. Y. Jiang, B. G. Lyeth, L. W. Jenkins, R. J. Hamm, and R. L. Hayes (1991). Marked protection by moderate hypothermia after experimental traumatic brain injury. *J. Cereb. Blood Flow Metab.,* 11:114–21.

Collins, J. C., T. C. Pilkington, and K. Schmidt-Nielsen (1971). A model of respiratory heat transfer in a small mammal. *Biophys. J.,* 11:886–914.

Collins, M. G., W. S. Hunter, and C. M. Blatteis (1987). Factors producing elevated core temperature in spontaneously hypertensive rats. *J. Appl. Physiol.,* 63:740–5.

Colquhoun, E. Q., and M. G. Clark (1991). Open question: Has thermogenesis in muscle been overlooked and misinterpreted? *News Physiol. Sci.,* 6:256–9.

Conklin, P., and F. W. Heggeness (1971). Maturation of temperature homeostasis in the rat. *Am. J. Physiol.,* 220:333–6.

Conn, C. A., K. T. Borer, and M. J. Kluger (1990). Body temperature rhythm and response to pyrogen in exercising and sedentary hamsters. *Med. Sci. Sports Exerc.,* 22:636–42.

Connolly, M. S., and C. B. Lynch (1981). Circadian variation of strain differences in body temperature and activity in mice. *Physiol. Behav.,* 27:1045–9.

Cook, L., C. J. Gordon, H. A. Tilson, and F. W. Edens (1987). Chlordecone-induced effects on thermoregulatory processes in the rat. *Toxicol. Appl. Pharmacol.,* 90:126–34.

Corbett, S. W., L. N. Kaufman, and R. E. Keesey (1988). Thermogenesis after lateral hypothalamic lesions: contributions of brown adipose tissue. *Am. J. Physiol.,* 255:E708–15.

Corbit, J. D. (1970). Behavioral regulation of body temperature. In J. D. Hardy, A. P. Gagge, and J. A. J. Stolwijk (eds.), *Physiological and Behavioral Temperature Regulation* (pp. 777–801). Springfield: Charles C Thomas.

Cowles, R. B., and C. M. Bogert (1944). A preliminary study of the thermal requirements of desert reptiles. *Bull. Am. Mus. Nat. Hist.,* 83:261–96.

Cox, B., T. Lee, and J. Parkes (1981). Decreased ability to cope with heat and cold linked to a dysfunction in a central dopaminergic pathway in elderly rats. *Life Sci.,* 28:2039–44.

Cox, B., and P. Lomax (1977). Pharmacological control of temperature regulation. *Annu. Rev. Pharmacol. Toxicol.,* 17:341–53.

Crawshaw, L. I. (1973). Effect of intracranial acetylcholine injection on thermoregulatory responses in the rat. *J. Comp. Physiol. Psychol.,* 83:32–5.

Cresteil, T. (1977). Noradrenaline appearance in guinea pig brown adipose tissue. *Biol. Neonate* 32:143–6.

Cui, J., G. Zaror-Behrens, and J. Himms-Hagen (1990). Capsaicin desensitization induces atrophy of brown adipose tissue in rats. *Am. J. Physiol.,* 259:R324–32.

Czaja, J. A., and P. C. Butera (1986). Body temperature and temperature gradients: changes during the estrous cycle and in response to ovarian steroids. *Physiol. Behav.,* 36:591–6.

Daniel, H., and D. M. Derry (1969). Criteria for differentiation of brown and white fat in the rat. *Can. J. Physiol. Pharmacol.,* 47:941–5.

Dascombe, M. J., N. J. Rothwell, B. O. Sagay, and M. J. Stock (1989). Pyrogenic and thermogenic effects of interleukin 1β in the rat. *Am. J. Physiol.*, 256:E7–11.

Dauncey, M. J. (1984). Behavioral and autonomic thermoregulation in lean and genetically obese (ob/ob) mice. *J. Therm. Biol.*, 9:247–53.

Dauncey, M. J., and D. Brown (1987). Role of activity-induced thermogenesis in twenty-four hour energy expenditure of lean and genetically obese (ob/ob) mice. *Q. J. Exp. Physiol.*, 72:549–59.

Davidson, I. W. F., J. C. Parker, and R. P. Beliles (1986). Biological basis for extrapolation across mammalian species. *Reg. Toxicol. Pharmacol.*, 6:211–37.

Davis, T. R. (1959). Thermogenic factors during cooling and in the stabilized hypothermic state. *Ann. N.Y. Acad. Sci.*, 80:500–14.

Davis, T. R., D. R. Johnston, F. C. Bell, and B. J. Cremer (1960). Regulation of shivering and non-shivering heat production during acclimation of rats. *Am. J. Physiol.*, 198:471–5.

Dawson, N. J., R. F. Hellon, J. G. Herington, and A. A. Young (1982). Facial thermal input in the caudal trigeminal nucleus of rats reared at 30°C. *J. Physiol.*, 333:545–54.

Dawson, N. J., and A. W. Kleber (1979). Physiology of heat loss from an extremity: the tail of the rat. *Clin. Exp. Pharmacol. Physiol.*, 6:69–80.

Dawson, N. J., and J. L. Malcolm (1981). The control of shivering in the rat. *Comp. Biochem. Physiol.*, 69A:43–9.

Deavers, D. R., and X. J. Musacchia (1979). The function of glucocorticoids in thermogenesis. *Fed. Proc.*, 38:2177–81.

De Luca, B., M. Monda, S. Amaro, M. P. Pellicano, and L. A. Cioffi (1989). Lack of diet-induced thermogenesis following lesions of paraventricular nucleus in rats. *Physiol. Behav.*, 46:685–91.

Demes, G. L., E. R. Buskirk, S. S. Alpert, and J. L. Loomis (1991). Energy turnover and heat exchange in mature lean and obese Zucker rats acutely exposed to three environmental temperatures for 24 hours. *Int. J. Obes.*, 15:375–85.

Depocas, F., J. S. Hart, and O. Héroux (1957). Energy metabolism of the white rat after acclimation to warm and cold environments. *J. Appl. Physiol.*, 10: 393–7.

Desautels, M., R. A. Dulos, and J. A. Thornhill (1985). Thermoregulatory responses of dystrophic hamsters to changes in ambient temperatures. *Can. J. Physiol. Pharmacol.*, 63:1145–50.

Dickenson, A. H. (1977). Specific responses of rat raphe neurones to skin temperature. *J. Physiol.*, 273:277–93.

Doi, K., and A. Kuroshima (1982a). Thermogenic response to glucagon in cold-acclimated mice. *Jpn. J. Physiol.*, 32:377–85.

(1982b). Sexual difference in the thermoregulatory ability of rats exposed to cold and heat. *J. Therm. Biol.*, 7:99–105.

(1984). Economy of hormonal requirement for metabolic temperature acclimation. *J. Therm. Biol.*, 9:87–91.

Doi, K., T. Ohno, and A. Kuroshima (1982). Role of endocrine pancreas in temperature acclimation. *Life Sci.*, 30:2253–9.

Doris, P. A., and M. A. Baker (1981). Hypothalamic control of thermoregulation during dehydration. *Brain Res.*, 206:219–22.

Doull, J. (1972). The effect of physical environmental factors on drug response. *Essays Toxicol.*, 3:37–63.

Duncan, W. C., B. Gao, and T. A. Wehr (1990). Light and antidepressant drugs: interactions with vigilance states, body temperature and oxygen consumption in Syrian hamsters. In J. Horne (ed.), *Sleep '90* (pp. 356–9). Bochum: Pontenagel country Press.

Dupré, R. K., and T. L. Owen (1992). Behavioral thermoregulation by hypoxic rats. *J. Exp. Zool.*, 262:230–5.

Dupré, R. K., A. M. Romero, and S. C. Wood (1988). Thermoregulation and metabolism in hypoxic animals. In N. C. Gonzalez and M. R. Fedde (eds.), *Oxygen Transfer from Atmosphere to Tissues* (pp. 347–51). New York: Plenum Press.

Durkot, M. J., R. P. Francesconi, and R. W. Hubbard (1986). Effect of age, weight, and metabolic rate on endurance, hyperthermia, and heat stroke mortality in a small animal model. *Aviat. Space Environ. Med.*, 57:974–9.

Eastman, C. I., and A. Rechtschaffen (1983). Circadian temperature and wake rhythms of rats exposed to prolonged continuous illumination. *Physiol. Behav.*, 31:417–27.

Eastman, C. I., R. E. Mistlberger, and A. Rechtschaffen (1984). Suprachiasmatic nuclei lesions eliminate circadian temperature and sleep rhythms in the rat. *Physiol. Behav.*, 32:357–68.

Edwards, M. J. (1969). Congenital defects in guinea pigs: prenatal retardation of brain growth of guinea pigs following hyperthermia during gestation. *Teratology*, 2:329–36.

(1982). The effects of hyperthermia on brain development. *Birth Defects*, 18(3A): 3–11.

Eedy, J. W., and D. M. Ogilvie (1970). The effect of age on the thermal preference of white mice (*Mus musculus*) and gerbils (*Meriones unguiculatus*). *Can. J. Zool.*, 48:1303–6.

Eide, R. (1976). Physiological and behavioral reactions to repeated tail cooling in white rat. *J. Appl. Physiol.*, 41:292–4.

Elliott, D. S., P. J. Burfening, and L. C. Ulberg (1968). Subsequent development during incubation of fertilized mouse ova stressed by high ambient temperatures. *J. Exp. Zool.*, 169:481–6.

Elmer, M., and P. Ohlin (1970). Salivary glands of the rat in a hot environment. *Acta Physiol. Scand.*, 79:129–32.

Else, P. L., and A. J. Hulbert (1987). Evolution of mammalian endothermic metabolism: "leaky" membranes as a source of heat. *Am. J. Physiol.*, 253:R1–7.

Epstein, A. N., and R. Milestone (1968). Showering as a coolant for rats exposed to heat. *Science*, 160:895–6.

Erskine, D. J., and V. H. Hutchison (1982a). Critical thermal maxima in small mammals. *J. Mammal.*, 63:267–73.

(1982b). The critical thermal maximum as a determinant of thermal tolerance in *Mus musculus*. *J. Therm. Biol.*, 7:125–31.

Estler, C.-J., and H. P. T. Ammon (1969). The importance of the adrenergic beta-receptors for thermogenesis and survival of acutely cold-exposed mice. *Can. J. Physiol. Pharmacol.*, 47:427–34.

Ettenberg, A., and H. J. Carlisle (1985). Neuroleptic-induced deficits in operant responding for temperature reinforcement. *Pharmacol. Biochem. Behav.*, 22:761–7.

Fantino, M., and M. Cabanac (1984). Effect of a cold ambient temperature on the rat's food hoarding behavior. *Physiol. Behav.*, 32:183–90.

Farkas, M. (1978). The role of the body mass:body surface ratio in thermoregulatory responses to cold, hypoxia, and hypercapnia in new-born, adult, and aged guinea pigs. *Experientia* [Suppl.], 32:297–301.

Feldberg, W., and R. D. Myers (1964). Effects on temperature of amines injected into the cerebral ventricles. A new concept of temperature regulation. *J. Physiol.*, 173:226–37.

Felice, R. T., and J. A. Miller, Jr. (1972). Hypothermia and resistance of mice to lethal exposure to high gravitational forces. *Aerospace Med.*, 43:860–8.

Ferguson, A. V., S. L. Turner, K. E. Cooper, and W. L. Veale (1985). Neurotransmitter effects on body temperature are modified with increasing age. *Physiol. Behav.*, 34:977–81.

Ferguson, A. V., W. I. Veale, and K. E. Cooper (1984). Changes in the hypothalamic mechanisms involved in the control of body temperature induced by early thermal environment. *Brain Res.*, 290:297–306.

Ferguson, J. H., and G. E. Folk, Jr. (1970). The critical thermal minimum of small rodents in hypothermia. *Cryobiology*, 7:44–6.

Ferguson, J. H., and T. D. Schultz (1975). Plasma free fatty acid composition before and after cold exposure in the white rat. *Int. J. Biochem.*, 6:69–72.

Finger, F. W. (1976). Relation of general activity in rats to environmental temperature. *Percept. Mot. Skills*, 43:875–90.

Flaim, K. E., J. M. Horowitz, and B. A. Horwitz (1976). Functional and anatomical characteristics of the nerve–brown adipose tissue interaction in the rat. *Pflügers Arch.*, 365:9–14.

Folk, G. E., Jr. (1974). *Textbook of Environmental Physiology*. Philadelphia: Lea & Febiger.

Ford, D. M., and K. P. Klugman (1980). Body mass and sex as determining factors in the development of fever in rats. *J. Physiol.*, 304:43–50.

Foster, D. O. (1984). Quantitative contribution of brown adipose tissue thermogenesis to overall metabolism. *Can. J. Biochem. Cell Biol.*, 62:618–22.

Foster, D. O., and M. L. Frydman (1978). Nonshivering thermogenesis in the rat. II: Measurements of blood flow with microspheres point to brown adipose tissue as the dominant site of the calorigenesis induced by noradrenaline. *Can. J. Physiol. Pharmacol.*, 56:110–22.

(1979). Tissue distribution of cold-induced thermogenesis in conscious warm- or cold-acclimated rats reevaluated from changes in tissue blood flow: the dominant role of brown adipose tissue in the replacement of shivering by nonshivering thermogenesis. *Can J. Physiol. Pharmacol.*, 57:257–70.

Fowler, S. J., and C. Kellog (1975). Ontogeny of thermoregulatory mechanisms in the rat. *J. Comp. Physiol. Psychol.*, 89:738–46.

Francesconi, R., R. Hubbard, and M. Mager (1984). Effects of pyridostigmine on ability of rats to work in the heat. *J. Appl. Physiol.*, 56:891–5.

Frankel, H. M. (1959). *Tolerance to high temperatures in small mammals* (128 p). Thesis. Microfilm 58-5819. Dissertation Abstracts, Jan. 1959.

Frankel, S., and G. Lange (1980). Maturation of hypothalamic-pituitary-thyroid response in the rat to acute cold. *Am. J. Physiol.*, 239:E223–6.

Franken, P., D. Dijk, I. Tobler, and A. A. Borbley (1991). Sleep deprivation in rats: effects on EEG power spectra, vigilance, and cortical temperature. *Am. J. Physiol.*, 261:R198–208.

Freeman, M.E., J. K. Crissman, Jr., G. N. Louw, R. L. Butcher, and E. K. Inskeep (1970). Thermogenic action of progesterone in the rat. *Endocrinology*, 86:717–20.

Fregly, M. J. (1956). Relationship between ambient temperature and the spontaneous running activity of normal and hypertensive rats. *Am. J. Physiol.*, 187: 297–301.

—— (1990). Activity of the hypothalamic-pituitary-thyroid axis during exposure to cold. In E. Schönbaum and P. Lomax (eds.), *Thermoregulation: Physiology and Biochemistry* (pp. 437–94). New York: Pergamon Press.

Fregly, M. J., K. M. Cook, and A.B. Otis (1963). Effect of hypothyroidism on tolerance of rats to heat. *Am. J. Physiol.*, 204:1039–44.

Fregly, M. J., D. C. Kikta, R. M. Threatte, J. L. Torres, and C. C. Barney (1989). Development of hypertension in rats during chronic exposure to cold. *J. Appl. Physiol.*, 66:741–9.

Fuhrman, G. J., and F. A. Fuhrman (1961). Effects of temperature on the action of drugs. *Annu. Rev. Pharmacol.*, 1:65–78.

Fujii, T., and Y. Ohtaki (1985). Sex-related hyperthermic response to chlorpromazine in the offspring of rats treated with imipramine. *Dev. Pharmacol. Ther.*, 8:364–73.

Fujinami, H., T. Komabayshi, T. Izawa, T. Nakamura, K. Suda, and M. Tsuboi (1991). In vivo adaptive control of β-receptors and adenylate cyclase during short-term heat exposure in rat parotid glands. *Comp. Biochem. Physiol.*, 98C:411–16.

Fuller, C. A., D. W. Griffin, and J. M. Horowitz (1991). Diurnal responses of mammals to acute exposure to a hyperdynamic environment. *Am. J. Physiol.*, 261:R842–7.

Fuller, C. A., B. A. Horwitz, and J. M. Horowitz (1975). Shivering and nonshivering thermogenic responses of cold-exposed rats to hypothalamic warming. *Am. J. Physiol.*, 228:1519–24.

Fuller, C. A., J. M. Horowitz, and B. A. Horwitz (1977). Effects of acceleration on thermoregulatory responses of unanesthetized rats. *J. Appl. Physiol.*, 42:74–9.

Furuyama, F. (1982). Strain difference in thermoregulation of rats surviving extreme heat. *J. Appl. Physiol.*, 52:410–15.

—— (1988). Thermal salivation and body water economics among Wistar rat strains. *Jpn. J. Vet. Sci.*, 50:415–23.

Furuyama, F., K. Ohara, and A. Ota (1984). Estimation of rat thermoregulatory ability based on body temperature response to heat. *J. Appl. Physiol.*, 57:1271–5.

Fyda, D. M., K. E. Cooper, and W. L. Veale (1991a). Contribution of brown adipose tissue to central PGE$_1$-evoked hyperthermia in rats. *Am. J. Physiol.*, 260:R59–66.

(1991b). Modulation of brown adipose tissue-mediated thermogenesis by lesions to the nucleus tractus solitarius in the rat. *Brain Res.*, 546:203–10.

Gallaher, E. J., D. A. Enger, and J. W. Swen (1985). Automated remote temperature measurement in small animals using a telemetry-microcomputer interface. *Comput. Biol. Med.*, 15:103–10.

Gambert, S. R., and J. J. Barboriak (1982). Effect of cold exposure on thyroid hormone in Fischer 344 rats of increasing age. *J. Gerontol.*, 37:684–7.

Ganong, W. F. (1975). *Review of Medical Physiology*, 7th ed. Los Altos: Lange Medical.

Gardner, C. A., and R. C. Webb (1986). Cold-induced vasodilation in isolated, perfused rat tail artery. *Am. J. Physiol.*, 251:H176–81.

Gautier, H., M. Bonora, S. B. M'Barek, and J. S. Sinclair (1991). Effects of hypoxia and cold acclimation on thermoregulation in the rat. *J. Appl. Physiol.*, 71:1355–63.

Gemmell, R. T., and J. R. S. Hales (1977). Cutaneous arteriovenous anastomoses in the tail but absent from the ear of the rat. *J. Anat.*, 124:355–8.

Georgiev, J. (1978). Influence of environmental conditions and handling on the temperature rhythm of the rat. *Biotelem. Patient Monitor.*, 5:229–34.

Germain, M., W. S. Webster, and M. J. Edwards (1985). Hyperthermia as a teratogen: parameters determining hyperthermia-induced head defects in the rats. *Teratology*, 31:265–72.

Giacchino, J., B. A. Horwitz, and J. M. Horowitz (1979). Thermoregulation in unrestrained rats during and after exposure to 1.5–4 *G*. *J. Appl. Physiol.*, 46:1049–53.

Gibbs, F. P. (1981). Temperature dependence of rat circadian pacemaker. *Am. J. Physiol.*, 241:R17–20.

Gilbert, T. M., and C. M. Blatteis (1977). Hypothalamic thermoregulatory pathways in the rat. *J. Appl. Physiol.*, 43:770–7.

Girardier, L. (1983). Brown fat and energy dissipating tissue. In L. Girardier and M. J. Stock (eds.), *Mammalian Thermogenesis* (pp. 50–98). London: Chapman & Hall.

Goodman, E. L., and J. P. Knochel (1991). Heat stroke and other forms of hyperthermia. In P. Mackowiak (ed.), *Fever: Basic Mechanisms and Management* (pp. 267–87). New York: Raven Press.

Goodrich, C. A. (1977). Measurement of body temperature in neonatal mice. *J. Appl. Physiol.*, 43:1102–5.

Gordon, C. J. (1983a). Influence of heating rate on control of heat loss from the tail in mice. *Am. J. Physiol.*, 244:R778–84.

(1983b). Behavioral and autonomic thermoregulation in mice exposed to microwave radiation. *J. Appl. Physiol.*, 55:1242–8.

(1983c). A review of terms and proposed nomenclature for regulated vs. forced, neurochemical induced changes in body temperature. *Life Sci.*, 32: 1285–95.

(1985). Relationship between autonomic and behavioral thermoregulation in the mouse. *Physiol. Behav.*, 34:687–90.

(1986). Relationship between behavioral and autonomic thermoregulation in the guinea pig. *Physiol. Behav.*, 38:827–31.

(1987). Relationship between preferred ambient temperature and autonomic thermoregulatory function in rat. *Am. J. Physiol.*, 252:R1130–7.

(1988). Simultaneous measurement of preferred ambient temperature and metabolism in rats. *Am. J. Physiol.*, 254:R229–34.

(1990a). Thermal biology of the laboratory rat. *Physiol. Behav.*, 47:963–91.

(1990b). Induction of regulated hypothermia in mice by urine administration. *J. Therm. Biol.*, 15:97–101.

(1991). Toxic-induced hypothermia and hypometabolism: Do they increase uncertainty in the extrapolation of toxicological data from experimental animals to humans? *Neurosci. Biobehav. Rev.*, 15:95–8.

(1993). Twenty-four hour rhythms of selected ambient temperature in rat and hamster. *Physiol. Behav.*, 53:257–63.

Gordon, C. J., K. S. Fehlner, and M. D. Long (1986). Relationship between autonomic and behavioral thermoregulation in the golden hamster. *Am. J. Physiol.*, 251:R320–4.

Gordon, C. J. and J. H. Ferguson (1980). The correlation between colonic cooling and survival time in the acute cold-exposed laboratory mouse: influence of cold acclimation. *J. Therm. Biol.*, 5:159–62.

Gordon, C. J., and L. Fogelson (1991a). Comparison of rats of the Fischer 344 and Long-Evans strains in their autonomic thermoregulatory response to trimethyltin administration. *J. Toxicol. Environ. Health*, 32:141–52.

(1991b). Comparative effects of hypoxia on behavioral thermoregulation in rats, hamsters, and mice. *Am. J. Physiol.*, 260:R120–5.

Gordon, C. J., and J. E. Heath (1980). Inhibition of running wheel activity in the hamster during chronic thermal stimulation of the anterior hypothalamus. *J. Therm. Biol.*, 5:103–5.

(1983). Reassessment of the neural control of body temperature: importance of oscillating neural and motor components. *Comp. Biochem. Physiol.*, 74A:479–89.

(1986). Integration and central processing in temperature regulation. *Annu. Rev. Physiol.*, 48:595–612.

Gordon, C. J., K. A. Lee, T. A. Chen, P. Killough, and J. S. Ali (1991). Dynamics of behavioral thermoregulation in the rat. *Am. J. Physiol.*, 261:R705–11.

Gordon, C. J., and M. D. Long (1984). Ventilatory frequency of mouse and hamster during microwave-induced heat exposure. *Respir. Physiol.*, 56:81–90.

Gordon, C. J., M. D. Long, and K. S. Fehlner (1984). Behavioural and autonomic thermoregulation in hamsters during microwave-induced heat exposure. *J. Therm. Biol.*, 9:271–7.

Gordon, C. J., and F. S. Mohler (1990). Thermoregulation at a high ambient temperature following the oral administration of ethanol in the rat. *Alcohol*, 7:551–5.

Gordon, C. J., A. H. Rezvani, M. E. Fruin, S. Trautwein, and J. E. Heath (1981). Rapid brain cooling in the free-running hamster *Mesocricetus auratus*. *J. Appl. Physiol.*, 51:1349–54.

Gordon, C. J., and A. G. Stead (1986). Effect of nickel and cadmium chloride on autonomic and behavioral thermoregulation in mice. *Neurotoxicology*, 7:97–106.

Gordon, C. J., and W. P. Watkinson (1988). Behavioral and autonomic thermoregulation in the rat following chlordimeform administration. *Neurotoxicol. Terat.*, 10:215–19.

Gordon, C. J., W. P. Watkinson, F. S. Mohler, and A. H. Rezvani (1988). Temperature regulation in laboratory mammals following acute toxic insult. *Toxicology,* 53:161–78.

Grant, M., and D. D. Thiessen (1989). The possible interaction of harderian material and saliva for thermoregulation in the Mongolian gerbil, *Meriones unguiculatus. Percept. Mot. Skills,* 68:3–10.

Grant, R. T. (1963). Vasodilation and body warming in the rat. *J. Physiol.,* 167:311–17.

Grayson, J., and D. Mendel (1956). The distribution and regulation of temperature in the rat. *J. Physiol.,* 133:334–46.

Griggio, M. A. (1982). The participation of shivering and nonshivering thermogenesis in warm and cold-acclimated rats. *Comp. Biochem. Physiol.,* 73A:481–4.

Grindeland, R. E., and G. E. Folk, Jr. (1962). Effects of cold exposure on the oestrous cycle of the golden hamster (*Mesocricetus auratus*). *J. Reprod. Fertil.,* 4:1–6.

Guernsey, D. L., and E. D. Stevens (1977). The cell membrane sodium pump as a mechanism for increasing thermogenesis during cold acclimation in rats. *Science,* 196:908–10.

Guernsey, D. L., and G. C. Whittow (1981). Basal metabolic rate, tissue thermogenesis and sodium-dependent tissue respiration of rats during cold-acclimation and deacclimation. *J. Therm. Biol.,* 6:7–10.

Günther, H., R. Brunner, and F. W. Klussmann (1983). Spectral analysis of tremorine and cold tremor electromyograms in animal species of different size. *Pflügers Arch.,* 399:180–5.

Guyton, A. C. (1986). *Textbook of Medical Physiology,* 7th ed. Philadelphia: Saunders.

Gwosdow, A. R., and E. L. Besch (1985). Effects of thermal history on the rat's response to varying environmental temperature. *J. Appl. Physiol.,* 59:413–19.

Gwosdow, A. R., E. L. Besch, C. L. Chen, and W. I. Li (1985). The effects of acclimation temperature on pituitary and plasma beta-endorphin in rats at 32.5°C. *Comp. Biochem. Physiol.,* 82C:269–72.

Habara, Y., and A. Kuroshima (1983). Changes in glucagon and insulin contents of brown adipose tissue after temperature acclimation in rats. *Jpn. J. Physiol.,* 33:661–5.

Habicht, G. S. (1981). Body temperature in normal and endotoxin-treated mice of different ages. *Mech. Ageing Dev.,* 16:97–104.

Hagerdal, M., F. A. Welsh, M. M. Keykhah, E. Perez, and J. R. Harp (1978). Protective effects of combinations of hypothermia and barbiturates in cerebral hypoxia in the rat. *Anesthesiology,* 49:165–9.

Hahn, P. (1956). Effect of environmental temperatures on the development of thermoregulatory mechanisms in infant rats. *Nature,* 178:96–7.

Hainsworth, F. R. (1967). Saliva spreading, activity, and body temperature regulation in the rat. *Am. J. Physiol.,* 212:1288–92.

(1968). Evaporative water loss from rats in the heat. *Am. J. Physiol.,* 214:979–82.

Hainsworth, F. R., and E. M. Stricker (1971). Evaporative cooling in the rat: differences between salivary glands as thermoregulatory effectors. *Can. J. Physiol. Pharmacol.,* 49:573–80.

Hainsworth, F. R., E. M. Stricker, and A. N. Epstein (1968). Water metabolism of rats in the heat: dehydration and drinking. *Am. J. Physiol.*, 214:983–9.

Hajos, M., and G. Engberg (1986). Emotional hyperthermia in spontaneously hypertensive rats. *Psychopharmacology,* 90:170–2.

Hammel, H. T., (1968). Regulation of internal body temperature. *Annu. Rev. Physiol.*, 30:641–710.

(1972). The set-point in temperature regulation: analogy or reality. In J. Bligh and R. E. Moore (eds.), *Essays on Temperature Regulation* (pp. 121–37). New York: Elsevier.

Hammel, H. T., D. C. Jackson, J. A. J. Stolwijk, J. D. Hardy, and S. B. Stomme (1963). Temperature regulation by hypothalamic proportional control with an adjustable set point. *J. Appl. Physiol.,* 18:1146–54.

Hansen, M. G., and I. Q. Wishaw (1973). The effects of 6-hydroxydopamine, dopamine and *dl*-norepinephrine on food intake and water consumption, selfstimulation, temperature and electroencephalographic activity in the rat. *Psychopharmacologia,* 29:33–44.

Harlow, H. J. (1987). Influence of the pineal gland and melatonin on blood flow and evaporative water loss during heat stress in rats. *J. Pineal Res.*, 4:147–59.

Harper, H. A. (1975). *Review of Physiological Chemistry,* 15th ed. Los Altos: Lange Medical.

Harri, M. N. E. (1976). Effect of body temperature on cardiotoxicity of isoprenaline in rats. *Acta Pharmacol. Toxicol.*, 39:214–24.

Harrison, G. A. (1958). The adaptability of mice to high environmental temperatures. *J. Exp. Biol.*, 35:892–901.

(1963). Temperature adaptation as evidenced by growth of mice. *Fed. Proc.*, 22:691–8.

Hart, J. S. (1951). Calorimetric determination of average body temperature of small mammals and its variation with environmental conditions. *Can. J. Zool.*, 29:224–33.

(1971). Rodents. In G. C. Whittow (ed.), *Comparative Physiology of Thermoregulation,* vol. 2 (pp. 1–149). New York: Academic Press.

Hart, J. S., O. Héroux, and F. Depocas (1956). Cold acclimation and the electromyogram of unanesthetized rats. *J. Appl. Physiol.*, 9:404–8.

Hart, J. S., and L. Janský (1963). Thermogenesis due to exercise and cold in warmand cold-acclimated rats. *Can. J. Biochem. Physiol.*, 41:629–34.

Hayashi, H., and T. Nakagawa (1963). Functional activity of the sweat glands of the albino rat. *J. Invest. Dermatol.*, 41:365–7.

Hayssen, V., and R. C. Lacy (1985). Basal metabolic rates in mammals: taxonomic differences in the allometery of BMR and body mass. *Comp. Biochem. Physiol.*, 81A:741–54.

Heath, J. E. (1986). Thermoregulation in vertebrates (introduction). *Annu. Rev. Physiol.*, 48:593–4.

Heath, J. E., B. A. Williams, and S. H. Mills (1971). Interactions of hypothalamic thermosensitivity in vertebrates. *Int. J. Biometeor.*, 15:254–7.

Heath, J. E., B. A. Williams, S. H. Mills, and M. J. Kluger (1972). The responsiveness of the preoptic–anterior hypothalamus to temperature in vertebrates. In F. E. South et al. (eds.), *Hibernation and Hypothermia, Perspectives and Challenges* (pp. 605–27). Amsterdam: Elsevier.

Heath, M. E. (1985). Effect of cutaneous denervation of face and trunk on thermoregulatory responses to cold in rats. *J. Appl. Physiol.*, 58:376–83.

Heldmaier, G. (1971). Relationship between non-shivering thermogenesis and body size. In L. Janský (ed.), *Non-shivering Thermogenesis. Proceedings of the Symposium* (pp. 73–80). Amsterdam: Swets & Zeitlinger N.V.

(1975). The influence of the social thermoregulation on the cold-adaptive growth of BAT in hairless and furred mice. *Pflügers Arch.*, 355:261–6.

Heller, H. C. (1978). Hypothalamic thermosensitivity in mammals. In L. Girardier and J. Seydoux (eds.), *Effectors of Thermogenesis* (pp. 267–76). Basel: Birkauser; *Experientia*, Suppl. 32.

(1979). Hibernation: neural aspects. *Annu. Rev. Physiol.*, 41:305–21.

Heller, H. C., and S. F. Glotzbach (1977). Thermoregulation during sleep and hibernation. *Int. Rev. Physiol., Environment. Physiol.*, 15:147–88.

Hellon, R. F. (1983). Central projections and processing of skin-temperature signals. *J. Therm. Biol.*, 8:7–8.

Hellon, R. F., H. Hensel, and K. Schäfer (1975). Thermal receptors in the scrotum of the rat. *J. Physiol.*, 248:349–57.

Hellon, R. F., and N. K. Misra (1973a). Neurones in the dorsal horn of the rat responding to scrotal skin temperature changes. *J. Physiol.*, 232:375–88.

(1973b). Neurones in the ventrobasal complex of the rat thalamus responding to scrotal skin temperature changes. *J. Physiol.*, 232:389–99.

Hellon, R. F., and D. C. M. Taylor (1982). An analysis of thermal afferent pathways in the rat. *J. Physiol.*, 326:319–28.

Hellström, B. (1975a). Heat vasodilation of the rat tail. *Can. J. Physiol. Pharmacol.*, 53:202–6.

(1975b). Cold vasodilation of the rat tail. *Can. J. Physiol. Pharmacol.*, 53:207–10.

Hensel, H. (1973). Neural processes in thermoregulation. *Physiol. Rev.*, 53:948–1016.

(1981). *Thermoreception and Temperature Regulation.* London: Academic Press.

Hensel, H., and K. Schäfer (1982). Dynamic and static activity of cold receptors in cats after long-term adaptation to various temperatures. *Pflügers Arch.*, 392:291–4.

Héroux, O. (1959). Comparison between seasonal and thermal acclimation in white rats. II: Surface temperature, vascularization, and in vitro respiration of the skin. *Can. J. Biochem. Physiol.*, 37:1247–53.

Héroux, O., F. Depocas, and J. S. Hart (1959). Comparison between seasonal and thermal acclimation in white rats. I: Metabolic and insulative changes. *Can. J. Biochem. Physiol.*, 37:473–8.

Héroux, O., E. Page, J. LeBlanc, J. Leduc, R. Gilbert, A. Villemaire, and P. Rivest (1975). Nonshivering thermogenesis and cold resistance in rats under severe cold conditions. *J. Appl. Physiol.*, 38:436–42.

Herrington, L. P. (1940). The heat regulation of small laboratory animals at various environmental temperatures. *Am. J. Physiol.*, 129:123–39.

Heusner, A. A. (1982). Energy metabolism and body size. I: Is the 0.75 mass exponent of Kleiber's equation a statistical artifact? *Respir. Physiol.*, 48:1–12.

Hill, R. M. (1947). The control of body temperature in white rats. *Am. J. Physiol.*, 149:650–6.

Himms-Hagen, J. (1984). Nonshivering thermogenesis. *Brain Res. Bull.*, 12:151–60.

(1986). Brown adipose tissue and cold-acclimation. In P. Trayhurn and D. G. Nicholls (eds.), *Brown Adipose Tissue* (pp. 214–68). London: Edward Arnold.

(1989). Brown adipose tissue thermogenesis and obesity. *Prog. Lipid Res.*, 28:67–115.

(1990a). Brown adipose tissue thermogenesis: role in thermoregulation, energy regulation and obesity. In E. Schönbaum and P. Lomax (eds.), *Thermoregulation: Physiology and Biochemistry* (pp. 327–414). New York: Pergamon Press.

(1990b). Brown adipose tissue thermogenesis: interdisciplinary studies. *FASEB J.*, 4:2890–3.

Himms-Hagen, J., J. Cui, and S. L. Sigurdson (1990). Sympathetic and sensory nerves in control of growth of brown adipose tissue: effects of denervation and of capsaicin. *Neurochem. Int.*, 17:271–9.

Himms-Hagen, J., and C. Gwilliam (1980). Abnormal brown adipose tissue in hamsters with muscular dystrophy. *Am. J. Physiol.*, 239:C18–22.

Hinckel, P., and W. T. Perschel (1987). Influence of cold and warm acclimation on neuronal responses in the lower brain stem. *Can. J. Physiol. Pharmacol.*, 65:1281–9.

Hinckel, P., and K. Schröder-Rosenstock (1981). Responses of pontine units to skin-temperature changes in the guinea-pig. *J. Physiol.*, 314:189–94.

Hirata, K., and T. Nagasaka (1981). Calorigenic and cardiovascular responses to norepinephrine in anesthetized and unanesthetized control and cold-acclimated rats. *Jpn. J. Physiol.*, 31:305–16.

Hirvonen, J., P. Huttunen, and H. Vapaatalo (1976). Serum glucose, serum free fatty acids and adipose tissue lipids after fatal hypothermia of cold acclimatized, reserpine or propranolol treated guinea-pigs. *Z. Rechtsmed.*, 77:177–89.

Hirvonen, J., D. Weaver, and D. D. Williams (1973). Morphological and enzyme histochemical changes in the interscapular adipose tissue of adult guinea-pigs during prolonged exposure to cold. *Experientia*, 29:1566–70.

Hissa, R., and K. Lagerspetz (1964). The postnatal development of homoiothermy in the golden hamster. *Ann. Med. Exp. Fenn.*, 42:43–5.

Hjeresen, D. L., and J. Diaz (1988). Ontogeny of susceptibility to experimental febrile seizures in rats. *Dev. Psychobiol.*, 21:261–75.

Hoffman, R. A. (1968). Hibernation and effects of low temperature. In R. A. Hoffman, P. F. Robinson, and H. Magalhaes (eds.), *The Golden Hamster, Its Biology and Use in Medical Research* (pp. 25–39). Ames: Iowa State University Press.

Hoffman-Goetz, L., and R. Keir (1984). Body temperature responses of aged mice to ambient temperature and humidity stress. *J. Gerontol.*, 39:547–51.

Hogan, S., and J. Himms-Hagen (1980). Abnormal brown adipose tissue in obese (ob/ob) mice: response to acclimation to cold. *Am. J. Physiol.*, 239:E301–9.

Holloway, B. R., R. G. Davidson, S. Freeman, H. Wheeler, and D. Stribling (1984). Post-natal development of interscapular (brown) adipose tissue in the guinea pig: effect of environmental temperature. *Int. J. Obes.* 8:295–303.

Holtzman, D., K. Obana, and J. Olson (1981). Hyperthermy-induced seizures in the rat pup: a model for febrile convulsions in children. *Science*, 213:1034–6.

Homburger, F. (1979). Myopathy of hamster dystrophy: history and morphologic aspects. *Ann. N.Y. Acad. Sci.*, 317:2–17.

Honma, K., and T. Hiroshige (1978). Simultaneous determination of circadian rhythms of locomotor activity and body temperature in the rat. *Jpn. J. Physiol.*, 28:159–69.

Hori, T. (1991). An update on thermosensitive neurons in the brain: from cellular biology to thermal and non-thermal homeostatic functions. *Jpn. J. Physiol.*, 41:1–22.

Horowitz, M. (1976). Acclimatization of rats to moderate heat: body water distribution and adaptability of the submaxillary gland. *Pflügers Arch.*, 366:173–6.

Horowitz, M., D. Argov, and R. Mizrahi (1983). Interrelationships between heat acclimation and salivary cooling mechanism in conscious rats. *Comp. Biochem. Physiol.*, 74A:945–9.

Horowitz, M., and N. Givol (1989). Heat acclimation and heat stress: cardiac output distribution, plasma volume expansion and the involvement of the adrenergic pathway. In P. Lomax and E. Schönbaum (eds.), *Thermoregulation: Research and Clinical Applications* (pp. 204–7). 7th International Symposium on the Pharmacology of Thermoregulation. Basel: Karger.

Horowitz, M., and U. Meiri (1985). Thermoregulatory activity in the rat: effects of hypohydration, hypovolemia and hypertonicity and their interaction with short-term heat acclimation. *Comp. Biochem. Physiol.*, 82A:577–82.

Horowitz, M., Y. Oron, and E. Atias (1978). Amylase activity, glycoprotein and electrolyte concentration in rat's submaxillary salivary gland during heat acclimation. *Comp. Biochem. Physiol.*, 60B:351–4.

Horowitz, M., Y. Shimoni, S. Parnes, M. S. Gotsman, and Y. Hasin (1986). Heat acclimation: cardiac performance of isolated rat heart. *J. Appl. Physiol.*, 60:9–13.

Horowitz, M., and W. A. Soskolne (1978). Cellular dynamics of rats' submaxillary gland during heat acclimatization. *J. Appl. Physiol.*, 44:21–4.

Horowitz, M., E. Sugimoto, T. Okuno, and T. Morimoto (1988). Changes in blood volume and vascular compliance during body heating in rats. *Pflügers Arch.*, 412:354–8.

Horwitz, B. A. (1978). Neurohumoral regulation of nonshivering thermogenesis in mammals. In L. C. H. Wang and J. W. Hudson (eds.), *Strategies in the Cold, Natural Torpidity and Thermogenesis* (pp. 619–53). New York: Academic Press, 1978.

(1979). Isoproterenol-induced calorigenesis of dystrophic and normal hamsters. *Proc. Soc. Exp. Biol. Med.*, 147:393–5.

Horwitz, B. A., and M. Eaton (1977). Oubain-sensitive liver and diaphragm respiration in cold-acclimated hamster. *J. Appl. Physiol.*, 42:150–3.

Horwitz, B. A., and G. E. Hanes (1974). Isoproterenol-induced calorigenesis of dystrophic and normal hamsters. *Proc. Soc. Exp. Biol. Med.*, 147:392–5.

(1976). Propranolol and pyrogen effects on shivering and nonshivering thermogenesis in rats. *Am. J. Physiol.*, 230:637–42.

Hošek, B., and J. Chlumecký (1967). Metabolic reaction and heat loss in hairless and normal mice during short-term adaptation to heat and cold. *Pflügers Arch.*, 296:248–55.

Houštěk, J., J. Koecký, Z. Rychter, and T. Soukup (1988). Uncoupling protein in embryonic brown adipose tissue–existence of nonthermogenic and thermogenic mitochondria. *Biochim. Biophys. Acta*, 935:19–25.

Hsieh, A. C. (1963). The basal metabolic rate of cold-adapted rats. *J. Physiol.*, 169:851–61.

Hsieh, A. C., N. Emery, and L. D. Carlson (1971). Calorigenic effect of norepinephrine in newborn rats. *Am. J. Physiol.*, 221:1568–71.

Hubbard, R. W., C. B. Matthew, and R. Francesconi (1982). Heat-stressed rat: effects of atropine, desalivation, or restraint. *J. Appl. Physiol.*, 53:1171–4.

Hubbard, R. W., W. T. Matthew, R. E. L. Criss, C. Kelly, I. Sils, M. Mager, W. D. Bowers, and D. Wolfe (1978). Role of physical effort in the etiology of rat heatstroke injury and mortality. *J. Appl. Physiol.*, 45:463–8.

Hutchison, V. H. (1980). The concept of critical thermal maximum. *Am. J. Physiol.*, 237:R367–8.

Huttunen, P., and J. Hirvonen (1977). The effect of ethanol on the ability of guineapigs to survive severe cold. *Forensic Sci.*, 9:185–93.

Huttunen, P., B. Kruk, M.-L. Kortelainen, V. Kinnula, and J. Hirvonen (1988). Physical training at low temperature promotes metabolic adaptation to cold in the guinea-pig. *J. Therm. Biol.*, 13:143–7.

Huttunen, P., H. Vapaatalo, and J. Hirvonen (1975). Catecholamine content in the interscapular adipose tissue and adrenal gland of cold-acclimatized guinea-pigs. *Acta Physiol. Scand.*, 93:574–6.

Imai-Matsumara, K., K. Matsumura, A. Morimoto, and T. Nakayama (1990). Suppression of cold-induced thermogenesis in full-term pregnant rats. *J. Physiol.*, 425:271–81.

Ishikawa, Y., T. Nakayama, K. Kanosue, and K. Matsumura (1984). Activation of central warm-sensitive neurons and the tail vasomotor response in rats during brain and scrotal thermal stimulation. *Pflügers Arch.*, 400:222–7.

Isler, D., P. Trayhurn, and P. G. Lunn (1984). Brown adipose tissue metabolism in lactating rats: the effect of litter size. *Ann. Nutr. Metab.*, 28:101–9.

Isobe, Y., S. Takaba, and K. Ohara (1980). Diurnal variation of thermal resistance in rats. *Can. J. Physiol. Pharmacol.*, 58:1174–9.

IUPS (1987). Glossary of terms for thermal physiology [revised by Committee on Thermal Physiology, International Union of Physiological Sciences (IUPS)]. *Pflügers Arch.*, 410:567–87.

Jakubczak, L. F. (1966). Behavioral thermoregulation in young and old rats. *J. Appl. Physiol.*, 21:19–21.

Jans, J. E., and M. Leon (1983). The effects of lactation and ambient temperature on the body temperature of female Norway rats. *Physiol. Behav.*, 30:959–61.

Janský, L. (1973). Non-shivering thermogenesis and its thermoregulatory significance. *Biol. Rev.*, 48:85–132.

(1978). Time sequence of physiological changes during hibernation: the significance of serotonergic pathways. In L. C. H. Wang and J. W. Hudson (eds.), *Strategies in Cold, Natural Torpidity and Thermogenesis* (pp. 299–326). New York: Academic Press.

(1979). Heat production. In P. Lomax and E. Schönbaum (eds.), *Body Temperature Regulation, Drug Effects, and Therapeutic Implications* (pp. 89–117). New York: Marcel Dekker.

(1990). Neuropeptides and the central regulation of body temperature during fever and hibernation. *J. Therm. Biol.*, 15:329–47.

Janský, L., and J. S. Hart (1968). Cardiac output and organ blood flow in warm- and cold-acclimated rats exposed to cold. *Can. J. Physiol. Pharmacol.*, 46:653–9.

Janssen, R. (1992). Thermal influences on nervous system function. *Neurosci. Biobehav. Rev.*, 16:399–413.

Jepson, M. M., D. J. Millward, N. J. Rothwell, and M. J. Stock (1988). Involvement of sympathetic nervous system and brown fat in endotoxin-induced fever in rats. *Am. J. Physiol.*, 255:E617–20.

Johanson, I. B. (1979). Thermotaxis in neonatal rat pups. *Physiol. Behav.*, 23:871–4.

Johnson, K. G., and M. Cabanac (1982). Homeostatic competition in rats fed at varying distances from a thermoneutral refuge. *Physiol. Behav.*, 29:7715–20.

Johnson, T. S., S. Murray, J. B. Young, and L. Landsberg (1982). Restricted food intake limits brown adipose tissue hypertrophy in cold exposure. *Life Sci.*, 30:1423–6.

Jones, S. B., and X. J. Musacchia (1976). Norepinephrine turnover in heart and spleen of 7-, 22-, and 34°C-acclimated hamsters. *Am. J. Physiol.*, 230:564–8.

Jones, S. B., X. J. Musacchia, and G. E. Temple (1976). Mechanisms of temperature regulation in heat-acclimated hamsters. *Am. J. Physiol.*, 231:707–12.

Joy, R. J. T., R. F. Knauft, and J. Mayer (1967). Simultaneous determination of regression equations for body composition measurements and metabolic rate in rat. *Proc. Soc. Exp. Biol. Med.*, 126:869–72.

Joy, R. J. T., and J. Mayer (1988). Caloric expenditure in cold-acclimating rats: an isogravimetric comparison. *Am. J. Physiol.*, 215:757–61.

Kalant, H., and A. D. Lê (1984). Effects of ethanol on thermoregulation. *Pharmacol. Ther.*, 23:313–64.

Kandasamy, S. B., and B. A. Williams (1983a). Opposing actions of dibutyrl cyclic AMP and GMP on temperature in conscious guinea-pigs. *Neuropharmacology*, 22:65–70.

(1983b). Hyperthermic effects of centrally injected (D-ALA^2, *N*-ME-PHE^4, MET-(*O*)^5-OL)-enkephalin (FK 33-824) in rabbits and guinea-pigs. *Neuropharmacology*, 22:1177–81.

(1984). Hypothermic and antipyretic effects of ACTH(1-24) and α-melanotropin in guinea-pigs. *Neuropharmacology*, 23:49–53.

Kanosue, K., T. Nakayama, Y. Ishikawa, and T. Hosono (1984). Threshold temperature of diencephalic neurons responding to scrotal warming. *Pflügers Arch.*, 400:418–23.

Kanosue, K., T. Nakayama, H. Tanaka, M. Yanase, and H. Yasuda (1990). Modes of action of local hypothalamic and skin thermal stimulation on salivary secretion in rats. *J. Physiol.*, 424:459–71.

Kanosue, K., K. Niwa, P. D. Andrew, H. Yasuda, M. Yanase, H. Tanaka, and K. Matsumura (1991). Lateral distribution of hypothalamic signal controlling thermoregulatory vasomotor activity and shivering in rats. *Am. J. Physiol.*, 260:R486–93.

Kant, G. J., R. A. Bauman, R. H. Pastel, C. A. Myatt, E. Closser-Gomez, and C. P. D'Angelo (1991). Effects of controllable vs. uncontrollable stress on circadian temperature rhythms. *Physiol. Behav.*, 49:625–30.

Kaplan, M. L., and G. A. Leveille (1974). Core temperature, O_2 consumption, and early detection of ob/ob genotype in mice. *Am. J. Physiol.*, 227:912–15.

Kaplanski, J., and Z. Ben-Zevi (1980). Effect of chronic heat exposure on in-vitro drug metabolism in the rat. *Life Sci.*, 26:639–42.

Kaplanski, J., E. Magal, U. A. Sod-Moriah, N. Hirschmann, and I. Nir (1983). The pineal and endocrine changes in heat exposed male hamsters. *J. Neural Transm.*, 58:261–70.

Kaplanski, J., R. Zohar, U. A. Sod-Moriah, E. Magal, N. Hirschmann, and I. Nir (1988). Pregnancy outcome in heat-exposed hamsters; the involvement of the pineal. *J. Neural Transm.*, 73:57–63.

Kasting, N. W. (1989). Criteria for establishing a physiological role for brain peptides. A case in point: the role of vasopressin in thermoregulation during fever and antipyresis. *Brain Res. Rev.*, 14:143–53.

Kasting, N. W., W. L. Veale, K. E. Cooper, and K. Lederis (1981). Vasopressin may mediate febrile convulsions. *Brain Res.*, 213:327–33.

Kates, A., and J. Himms-Hagen (1990). Defective regulation of thyroxine 5'-deiodinase in brown adipose tissue of ob/ob mice. *Am. J. Physiol.*, 258:E7–15.

Kennedy, W. R., M. Sakuta, and D. C. Quick (1984). Rodent eccrine sweat glands: a case of multiple efferent innervation. *Neuroscience*, 11:741–9.

Kent, S., M. Hurd, and E. Satinoff (1991). Interactions between body temperature and wheel running over the estrous cycle in rats. *Physiol. Behav.*, 49:1079–84.

Keplinger, M. L., G. E. Lanier, and W. B. Deichmann (1959). Effects of environmental temperature on the acute toxicity of a number of compounds in rats. *Toxicol. Appl. Pharmacol.*, 1:156–61.

Kerr, J. S., R. L. Squibb, and H. M. Frankel (1975). Effect of heat acclimation (32°C) on rat liver and brain substrate levels. *Int. J. Biochem.*, 6:191–5.

Kiang-Ulrich, M., and S. M. Horvath (1984a). Age-related metabolic modifications in male F344 rats. *Exp. Ageing Res.*, 10:89–93.

  (1984b). Metabolic responses to tyramine and cold in young male Sprague-Dawley and Fischer 344 rats. *Am. J. Physiol.*, 246:E141–4.

  (1985). Age-related differences in response to acute cold challenge (-10°C) in male F344 rats. *Exp. Gerontol.*, 20:201–9.

Kilham, L., and V. H. Ferm (1976). Exencephaly in fetal hamsters following exposure to hyperthermia. *Teratology*, 14:323–6.

Kittrell, E. M., and E. Satinoff (1986). Development of the circadian rhythm of body temperature in rats. *Physiol. Behav.*, 38:99–104.

  (1988). Diurnal rhythms of body temperature, drinking and activity over reproductive cycles. *Physiol. Behav.*, 42:477–84.

Kleiber, M. (1961). *The Fire of Life. An Introduction to Animal Energetics*. New York: Wiley.

  (1972a). A new Newton's Law of Cooling. *Science*, 178:1283–5.

  (1972b). Body size, conductance for animal heat flow and Newton's law of cooling. *J. Theor. Biol.*, 37:139–50.

Kleiber, M., and H. H. Cole (1950). Body size, growth rate, and metabolic rate in two inbred strains of rats. *Am. J. Physiol.*, 161:294–9.

Kleinebeckel, D., and F. W. Klussmann, (1990). Shivering. In E. Schonbaum and P. Lomax (eds.), *Thermoregulation: Physiology and Biochemistry* (pp. 235–53). New York: Pergamon Press.

Kleitman, N., and E. Satinoff (1982). Thermoregulatory behavior in rat pups from birth to weaning. *Physiol. Behav.*, 29:537–41.

Klir, J. J., J. E. Heath, and N. Bennani (1990). An infrared thermographic study of surface temperature in relation to external thermal stress in the Mongolian gerbil, *Meriones unguiculatus. Comp. Biochem. Physiol.*, 96A:141–6.

Kloog, Y., M. Horowitz, U. Meiri, R. Galron, and A. Avron (1985). Regulation of submaxillary gland muscarinic receptors during heat acclimation. *Biochim. Biophys. Acta*, 845:428–35.

Kluger, M. J. (1991). Fever: role of pyrogens and cryogens. *Physiol. Rev.*, 71:93–127.

Kluger, M. J., C. A. Conn, B. Franklin, R. Freter, and G. D. Abrams (1990). Effect of gastrointestinal flora on body temperature of rats and mice. *Am. J. Physiol.*, 258:R552–7.

Kluger, M. J., B. O'Reilly, T. R. Shope, and A. J. Vander (1987). Further evidence that stress hyperthermia is a fever. *Physiol. Behav.*, 39:763–6.

Knecht, E. A., M. A. Toraason, and G. L. Wright (1980). Thermoregulatory ability of female rats during pregnancy and lactation. *Am. J. Physiol.*, 239:R470–5.

Knox, G. V., C. Campbell, and P. Lomax (1973). Cutaneous temperature and unit activity in the hypothalamic thermoregulatory centers. *Exp. Neurol.*, 40:717–30.

Kodama, A. M., and N. Pace (1964). Effect of environmental temperature on hamster body fat composition. *J. Appl. Physiol.*, 19:863–7.

Kopecky, J., L. Sigurdson, I. R. Park, and J. Himms-Hagen (1986). Thyroxine 5′-deiodinase in hamster and rat brown adipose tissue: effect of cold and diet. *Am. J. Physiol.*, 251:E1–7.

Kortelainen, M.-L., P. Huttunen, and T. Lapinlampi (1990). Influence of two β-adrenoceptor antagonists, propranolol and pindolol, on cold adaptation in the rat. *Br. J. Pharmacol.*, 99:673–8.

Kotby, S., and H. D. Johnson (1967). Rat adrenal cortical activity during exposure to a high (34°C) ambient temperature. *Life Sci.*, 6:1121–32.

Kregel, K. C., and C. V. Gisolfi (1990). Circulatory responses to vasoconstrictor agents during passive heating in the rat. *J. Appl. Physiol.*, 68:1220–7.

Kregel, K. C., P. T. Wall, and C. V. Gisolfi (1988). Peripheral vascular responses to hyperthermia in the rat. *J. Appl. Physiol.*, 64:2582–8.

Krog, H., M. Monson, and L. Irving (1955). Influence of cold upon the metabolism and body temperature of wild rats, albino rats and albino rats conditioned to cold. *J. Appl. Physiol.*, 7:349–54.

Kuhnen, G. (1986). O_2 and CO_2 concentrations in burrows of euthermic and hibernating golden hamsters. *Comp. Biochem. Physiol.*, 84A:517–22.

Kurosawa, M. (1991). Reflex changes in thermogenesis in the interscapular brown adipose tissue in response to thermal stimulation of the skin via sympathetic efferent nerves in anesthetized rats. *J. Auton. Nerv. Syst.*, 33:15–24.

Kuroshima, A., and T. Ohno (1991). Cold- and noradrenaline-induced changes in ganglioside GM_3 levels of rat brown adipose tissue. *J. Therm. Biol.*, 16:37–9.

Kuroshima, A., and T. Yahata (1985). Effect of food restriction on cold adaptability of rats. *Can. J. Physiol. Pharmacol.*, 63:68–71.

Kuroshima, A., T. Yahata, K. Doi, and T. Ohno (1982). Thermal and metabolic responses of temperature-acclimated rats during cold and heat exposure. *Jpn. J. Physiol.*, 32:561–71.

Kuroshima, A., T. Yahata, and T. Ohno (1991). Comparison of *in vitro* thermogenesis of brown adipose tissue in cold-acclimated rats and guinea pigs. *J. Therm. Biol.*, 16:109–14.

Lackey, W. W., L. A. Broome, J. A. Goetting, and D. A. Vaughan (1970). Diurnal patterns of rats determined by calorimetry under controlled conditions. *J. Appl. Physiol.*, 29:824–9.

Lagerspetz, K. Y. (1966). Postnatal development of thermoregulation in laboratory mice. *Helg. Wiss. Meeresunters.*, 14:559–71.

Landsberg, L., M. E. Saville, and J. B. Young (1984). Sympathoadrenal system and regulation of thermogenesis. *Am. J. Physiol.*, 247:E181–9.

Lary, J. M., D. L. Conover, P. H. Johnson, and R. W. Hornung (1986). Dose-response relationship between body temperature and birth defects in radiofrequency-irradiated rats. *Bioelectromagnetics*, 7:141–9.

Laties, V. G., and B. Weiss (1960). Behavior in the cold after acclimatization. *Science*, 131:1891–2.

Laudenslager, M. L., C. W. Wilkinson, H. J. Carlisle, and H. T. Hammel (1980). Energy balance in ovariectomized rats with and without estrogen replacement. *Am. J. Physiol.*, 238:R400–5.

Laughter, J. S., Jr., and C. M. Blatteis (1985). A system for the study of behavioral thermoregulation of small animals. *Physiol. Behav.*, 35:993–7.

Lee, T. F., and L. C. H. Wang (1985). Improving cold tolerance in elderly rats by aminophylline. *Life Sci.*, 36:2025–32.

Leon, M. (1985). Development of thermoregulation. In E. M. Blass (ed.), *Handbook of Behavioral Neurobiology, Vol. 8* (pp. 297–322). New York: Plenum.

Leonard, C. M. (1974). Thermotaxis in golden hamster pups. *J. Comp. Physiol. Psychol.*, 86:458–69.

  (1982). Shifting strategies for behavioral thermoregulation in developing golden hamsters. *J. Comp. Physiol. Psychol.*, 96:234–43.

Leung, P. M., and B. A. Horwitz (1976). Free-feeding patterns of rats in response to changes in environmental temperature. *Am. J. Physiol.*, 231:1220–4.

Lewis, R. J., and R. L. Totken (eds.) (1979). *Registry of Toxic Effects of Chemical Substances.* U.S. Department of Health and Human Services Publ. No. 80-111.

Lin, M. T., and C. Y. Chai (1974). Independence of spinal cord and medulla oblongata on thermal activity. *Am. J. Physiol.*, 226:1066–72.

Lin, M. T., Y. F. Chern, G. G. Liu, and T. C. Chang (1979). Studies on thermoregulation in the rat. *Proc. Natl. Sci. Counc. ROC*, 3:46–52.

Lin, M. T., H. C. Wang, and A. Chandra (1980). The effects on thermoregulation of intracerebroventricular injections of acetylcholine, pilocarpine, physostigmine, atropine and hemicholinium in the rat. *Neuropharmacology*, 19:561–5.

Lin, M. T., S. J. Won, J. Fan, and C. Y. Chai (1990). Paramedian reticular nucleus: a thermolytic area in the rat medulla oblongata. *Pflügers Arch.*, 417:418–24.

Linder, M. D., and V. K. Gribkoff (1991). Relationship between performance in the Morris water task, visual acuity, and thermoregulatory function in aged F-344 rats. *Behav. Brain Res.*, 45:45–55.

Lipton, J. M. (1968). Effects of preoptic lesions on heat-escape responding and colonic temperature in the rat. *Physiol. Behav.*, 3:165–9.

  (1973). Thermosensitivity of medulla oblongata in control of body temperature. *Am. J. Physiol.*, 224:890–7.

(1989). Neuropeptide α-melanocyte-stimulating hormone in control of fever, the acute phase response, and inflamation. In *Neuroimmune Networks: Physiology and Diseases* (pp. 243–50). New York: Alan R. Liss.

Lipton, J. M., and W. G. Clark (1986). Neurotransmitters in temperature control. *Annu. Rev. Physiol.*, 48:613–23.

Lipton, J. M., P. E. Dwyer, and D. E. Fossler (1974). Effects of brainstem lesions on temperature regulation in hot and cold environments. *Am. J. Physiol.*, 226:1356–65.

Lomax, E. (1966). Measurement of core temperature in the rat. *Nature*, 210:854–5.

(1979). Historical development of concepts of thermoregulation. In P. Lomax and E. Schönbaum (eds.), *Body Temperature: Regulation, Drug Effects, and Therapeutic Implications* (pp. 1–23). New York: Marcel Dekker.

Lomax, P., and R. J. Lee (1982). Cold acclimation and resistance to ethanol-induced hypothermia. *Eur. J. Pharmacol.*, 84:87–91.

Long, N. C., A. J. Vander, and M. J. Kluger (1990). Stress-induced rise of body temperature in rats is the same in warm and cool environments. *Physiol. Behav.*, 47:773–5.

Lotz, W. G., and S. M. Michaelson (1978). Temperature and corticosterone relationships in microwave-exposed rats. *J. Appl. Physiol.*, 44:438–45.

Lovegrove, B. G., G. Heldmaier, and T. Ruf (1991). Perspectives of endothermy revisited: the endothermic temperature range. *J. Therm. Biol.*, 16:185–97.

Luebbert, S. J., L. E. McGregor, and J. C. Roberts (1979). Temperature acclimation in the Mongolian gerbil (*Meriones unguiculatus*): changes in metabolic rate and the response to norepinephrine. *Comp. Biochem. Physiol.*, 63A:169–75.

Lyman, C. P. (1948). The oxygen consumption and temperature regulation of hibernating hamsters. *J. Exp. Zool.*, 109:55–78.

(1990). Pharmacological aspects of mammalian hibernation. In E. Schönbaum and P. Lomax (eds.), *Thermoregulation: Physiology and Biochemistry* (pp. 415–36). New York: Pergamon Press.

Ma, S. W. Y., and D. O. Foster (1989). Brown adipose tissue, liver, and diet-induced thermogenesis in cafeteria diet-fed rats. *Can. J. Physiol. Pharmacol.*, 67:376–81.

McArthur, A. J. (1991). Metabolism of homeotherms in the cold and estimation of thermal insulation. *J. Therm. Biol.*, 16:149–55.

McCarty, R. (1985). Sympathetic-adrenal medullary and cardiovascular responses to acute cold stress in adult and aged rats. *J. Auton. Nerv. Syst.*, 12:15–22.

McCaughran, J. A., Jr., and N. Schechter (1982). Experimental febrile convulsions: long-term effects of hyperthermia-induced convulsions in the developing rat. *Epilepsia*, 23:173–83.

McDonald, R. B., C. Day, K. Carlson, J. S. Stern, and B. A. Horwitz (1989). Effect of age and gender on thermoregulation. *Am. J. Physiol.*, 257:R700–4.

McDonald, R. B., B. A. Horwitz, J. S. Hamilton, and J. S. Stern (1988a). Cold- and norepinephrine-induced thermogenesis in younger and older Fischer 344 rats. *Am. J. Physiol.*, 254:R457–62.

McDonald, R. B., B. A. Horwitz, and J. S. Stern (1988b). Cold-induced thermogenesis in younger and older Fischer 344 rats following exercise training. *Am. J. Physiol.*, 254:R908–16.

MacFarlane, B. A., and A. N. Epstein (1981). Biobehavioral determinants of evaporative water loss in the rat. *Behav. Neur. Biol.*, 33:101–16.

MacKintosh, D. A., J. Baird-Lambert, and N. Buchanan (1984). Changes in response to metrazole during fever in juvenile rats; a new model for febrile convulsions? *Acta Neurol. Scand.*, 69:147–53.

McManus, J. J. (1971). Early postnatal growth and the development of temperature regulation in the Mongolian gerbil, *Meriones unguiculatus. J. Mammal.*, 52:782–92.

McNab, B. (1980). On estimating thermal conductance in endotherms. *Physiol. Zool.*, 53:145–56.

McRae, M. A., and J. L. Hanegan (1981). The effect of para-chlorophenylalanine on thermoregulation in the Mongolian gerbil (*Meriones unguiculatus*). *Comp. Biochem. Physiol.*, 68C:181–6.

Magal, E., J. Kaplanski, U. A. Sod-Moriah, N. Hirschmann, and I. Nir (1981). Role of the pineal gland in male rats chronically exposed to increased temperature. *J. Neural Transm.*, 50:267–73.

  (1983). Effect of pinealectomy on heat-induced endocrine changes in female rats. *Horm. Res.*, 17:222–7.

Magnen, J. L. (1983). Body energy balance and food intake: a neuroendocrine regulatory mechanism. *Physiol. Rev.*, 63:314–86.

Maickel, R. P., N. Matussek, D. N. Stern, and B. B. Brodie (1967). The sympathetic nervous system as a homeostatic mechanism. I: Absolute need for sympathetic nervous function in body temperature maintenance of cold-exposed rats. *J. Pharmacol. Exp. Ther.*, 157:103–10.

Manning, J. M., and F. H. Bronson (1990). The effects of low temperature and food intake on ovulation in domestic mice. *Physiol. Zool.*, 63:938–48.

Markley, K., E. Smallman, S. W. Thornton, and G. Evans (1973). The effect of environmental temperature and fluid therapy on mortality and metabolism of mice after burn and tourniquet trauma. *J. Trauma*, 13:145–60.

Marques, P. R., R. L. Spencer, T. F. Burks, and J. N. McDougal (1984). Behavioral thermoregulation, core temperature, and motor activity: simultaneous quantitative assessment in rats after dopamine and prostaglandin E_1. *Behav. Neurosci.*, 98:858–67.

Marrone, B. L., R. T. Gentry, and G. N. Wade (1976). Gonadal hormones and body temperature in rats: effects of estrous cycles, castration and steroid replacement. *Physiol. Behav.*, 17:419–25.

Martin, J. R., A. Fuchs, and J. Harting (1985). Senescent rats of two strains exhibit greater cold-induced hypothermia than adult controls. *IRCS Med. Sci.*, 13:55–6.

Matthew, C. B., R. W. Hubbard, R. Francesconi, and P. C. Szlyk (1986). An atropinized heat-stressed rat model: dose response effects and pharmacokinetics. *Aviat. Space Environ. Med.*, 57:659–63.

Mautz, W. J., and C. Bufalino (1989). Breathing pattern and metabolic rate responses of rats exposed to ozone. *Respir. Physiol.*, 76:69–78.

Maxson, S. C. (1980). Febrile convulsions in inbred stains of mice susceptible and resistant to audiogenic seizures. *Epilepsia*, 21:637–45.

Mercer, J. B., and E. Simon (1983). A comparison between total body thermosensitivity and local thermosensitivity in the guinea pig (*Cavia porcellus*). *J. Therm. Biol.*, 8:43–5.

_____ (1984). A comparison between total body thermosensitivity and local thermosensitivity in mammals and birds. *Pflügers Arch.*, 400:228–34.

Merker, G., J. Roth, and E. Zeisberger (1989). Thermoadaptive influence on reactivity pattern of vasopressinergic neurons in the guinea pig. *Experientia*, 45:722–6.

Milner, R. E., and P. Trayhurn (1990). Rapid quantitation of uncoupling protein in brown adipose tissue mitochondria by dot immunobinding (''dot blot'') procedure: application to the measurement of uncoupling protein in Richardson's ground squirrel, rats, and mice. *Biochem. Cell. Biol.*, 68:973–9.

Minard, F. N., and D. S. Grant (1982). Hypothermia as a mechanism for drug-induced resistance to hypoxia. *Biochem. Pharmacol.*, 31:1197–203.

Monteith, J. L., and L. E. Mount (eds.) (1974). *Heat Loss from Animals and Man*. London: Butterworth.

Monteith, J. L., and M. H. Unsworth (eds.) (1990). *Principles of Environmental Physics*. London: Edward Arnold.

Moore, R. E., and B. Donne (1984). Nonshivering thermogenesis in the newborn mouse. In J. R. S. Hales (ed.), *Thermal Physiology* (pp. 175–8). New York: Raven Press.

Morimoto, A., N. Murakami, T. Nakamori, and T. Watanabe (1986). Suppression of nonshivering thermogenesis in the rat by heat-seeking behavior during cold exposure. *J. Physiol.*, 380:541–9.

Morimoto, A., N. Murakami, Y. Sakata, T. Watanabe, and K. Yamaguchi (1990). Functional and structural differences in febrile mechanism between rabbits and rats. *J. Physiol.*, 427:227–39.

Morley, R. M., C. A. Conn, M. J. Kluger, and A. J. Vander (1990). Temperature regulation in biotelemetered spontaneously hypertensive rats. *Am. J. Physiol.*, 258:R1064–9.

Morrison, P. R. (1948). Oxygen consumption in several mammals under basal conditions. *J. Cell. Comp. Physiol.*, 31:281–91.

Morrison, S. D. (1968). The constancy of the energy expended by rats on spontaneous activity, and the distribution of activity between feeding and non-feeding. *J. Physiol.*, 197:305–23.

Mount, L. E. (1971). Metabolic rate and thermal insulation in albino and hairless mice. *J. Physiol.*, 217:315–26.

Mouroux, I., R. Bertin, and R. Portet (1990). Thermogenic capacity of the brown adipose tissue of development rats; effects of rearing temperature. *J. Dev. Physiol.*, 14:337–42.

Muraki, T., and R. Kato (1987). Genetic analysis of hypothermia induced by morphine in two strains of inbred mice. *Pharmacol. Biochem. Behav.*, 27:87–91.

Musacchia, X. J., and M. Jacobs (1973). Helium-cold induced hypothermia in the white rat. *Proc. Soc. Exp. Biol. Med.*, 142:734–9.

Myer, J. L., I. Van Kersen, B. Becker, and G. M. Hahn (1985). The significance of thermotolerance after 41°C hyperthermia: in vivo and in vitro tumor and normal tissue investigations. *Int. J. Radiat. Oncol. Biol. Phys.*, 11:973–81.

Myers, R. D. (1980). Hypothalamic control of thermoregulation. In P. J. Morgane and J. Panksepp (eds.), *Handbook of the Hypothalamus*, vol. 3 (pp. 83–210). New York: Marcel Dekker.

(1981). Alcohol's effect on body temperature: hypothermia, hyperthermia or poikilothermia? *Brain Res. Bull.*, 7:209–20.

(1982). The role of ions in thermoregulation and fever. In A. S. Milton (ed.), *Handbook of Experimental Pharmacology*, vol. 60 (pp. 151–86). Berlin: Springer-Verlag.

(1987). Cholinergic systems in the central control of body temperature. In N. J. Dun and R. L. Perlman (eds.), *Neurobiology of Acetylcholine* (pp. 391–402). New York: Plenum Press.

Myers, R. D., and T. F Lee (1989). Neurochemical aspects of thermoregulation. In L. C. H. Wang (ed.), *Advances in Comparative and Environmental Physiology*, vol. 4 (pp. 161–203). Berlin: Springer-Verlag.

Myers, R. D., and M. B. Waller (1977). Thermoregulation and serotonin. In W. B. Essman (ed.), *Serotonin in Health and Disease. Vol. 2: Physiological Regulation and Pharmacological Action* (pp. 1–67). New York: Spectrum.

Myers, R. D., and T. L. Yaksh (1968). Feeding and temperature responses in the unrestrained rat after injections of cholinergic and aminergic substances into the cerebral ventricles. *Physiol. Behav.*, 3:917–28.

Nagasaka, T., K. Hirata, Y. Sugano, and H. Shibata (1979). Heat balance during physical restraint in rats. *Jpn. J. Physiol.*, 29:383–92.

Nagel, J. A., and E. Satinoff (1980). Mild cold exposure increases survival in rats with medial preoptic lesions. *Science*, 208:301–3.

Nakatsuka, H., Y. Shoji, and T. Tsuda (1983). Effects of cold exposure on gaseous metabolism and body composition in the rat. *Comp. Biochem. Physiol.*, 75A:21–5.

Nakayama, T. (1985). Thermosensitive neurons in the brain. *Jpn. J. Physiol.*, 35:375–89.

Nakayama, T., Y. Ishikawa, and T. Tsuratani (1979). Projection of scrotal thermal afferents to the preoptic and hypothalamic neurons in rats. *Pflügers Arch.*, 380:59–64.

Nakayama, T., K. Kanosue, Y. Ishikawa, K. Matsumura, and K. Imai (1983). Dynamic response of preoptic and hypothalamic neurons to scrotal thermal stimulation. *Pflügers Arch.*, 396:23–6.

Nakayama, T., K. Kanosue, H. Tanaka, and T. Kaminaga (1986). Thermally induced salivary secretion in anesthetized rats. *Pflügers Arch.*, 406:351–5.

Nattie, E. E., and J. E. Melton (1979). Breathing in the potassium depleted rat: the role of metabolic rate and body temperature. *Respir. Physiol.*, 38:223–33.

Nava, M. P., M. Ablenda, and M. L. Puerta (1990). Cold-induced and diet-induced thermogenesis in progesterone-treated rats. *Pflügers Arch.*, 415:747–50.

Nechad, M. (1986). Structure and development of brown adipose tissue. In P. Trayhurn and D. G. Nicholls (eds.), *Brown Adipose Tissue* (pp. 1–30). London: Edward Arnold.

Neya, T., and F.-K. Pierau (1976). Vasomotor responses to thermal stimulation of the scrotal skin in rats. *Pflügers Arch.*, 363:15–18.

(1980). Activity patterns of temperature-reactive dorsal horn neurons and their reactions to peripheral receptor stimulation by Ca. *Jpn. J. Physiol.*, 30:921–34.

Nicholls, D. G., and R. M. Locke (1984). Thermogenic mechanisms in brown fat. *Physiol. Rev.*, 64:1–64.

Nir, I., and N. Hirschmann (1978). Pineal *N*-acetyltransferase depression in rats exposed to heat. *Experientia*, 34:1645–7.

Nomiyama, K., K. Matsui, and H. Nomiyama (1980a). Effects of temperature and other factors on the toxicity of methylmercury in mice. *Toxicol. Appl. Pharmacol.*, 56:392–8.

—— (1980b). Environmental temperature, a factor modifying the acute toxicity of organic solvents, heavy metals, and agricultural chemicals. *Toxicol. Lett.*, 6:67–70.

Nuesslein, B., and I. Schmidt (1990). Development of circadian cycle of core temperature in juvenile rats. *Am. J. Physiol.*, 259:R270–6.

Obál, F., G. Rubicsek, P. Alföldi, G. Sáry, and F. Obál (1985). Changes in the brain and core temperatures in relation to the various arousal states in rats in the light and dark periods of the day. *Pflügers Arch.*, 404:73–9.

Ogilvie, D. M., and R. H. Stinson (1966). The effect of age on temperature selection by laboratory mice (*Mus musculus*). *Can. J. Zool.*, 44:511–17.

Ohara, K., F. Furuyama, and Y. Isobe (1975). Prediction of survival time of rats in severe heat. *J. Appl. Physiol.*, 38:724–9.

Ohara, K., H. Sato, N. Okuda, Y. Makino, and Y. Isobe (1982). Response in rectal and skin temperatures to centrifugal forces in rats of different ambient temperatures. *Int. J. Biometeor.*, 26:61–72.

Ohno, T., and A. Kuroshima (1986). Muscle myoglobin as determined by electrophoresis in thermally acclimated rat. *Jpn. J. Physiol.*, 36:733–44.

Ohno, T., T. Yahata, and A. Kuroshima (1990). Changes in fasting-induced adrenocortical secretion of cold-acclimated rats. *Jpn. J. Physiol.*, 40:463–70.

Ohtake, M., G. A. Bray, and M. Azukizawa (1977). Studies on hypothermia and thyroid function in the obese (ob/ob) mouse. *Am. J. Physiol.*, 233:R110–15.

Okamoto, K., and K. Aoki (1963). Development of a strain of spontaneously hypertensive rats. *Jpn. Circ. J.*, 27:282–93.

Om, P., and M. Hohenegger (1980). Energy metabolism in acute uremic rats. *Nephron*, 25:249–53.

Oufara, S., H. Barre, J.-L. Rouanet, and J. Chatonnet (1987). Adaptation to extreme ambient temperatures in gerbils and mice. *Am. J. Physiol.*, 253:R39–45.

Overstreet, D. H., R. W. Russel, A. D. Crocker, J. C. Gillin, and D. S. Janowsky (1988). Genetic and pharmacological models of cholinergic supersensitivity and affective disorders. *Experientia*, 44:465–72.

Owen, T. L., R. L. Spencer, and S. P. Duckles (1991). Effect of age on cold acclimation in rats: metabolic and behavioral responses. *Am. J. Physiol.*, 260:R284–9.

Oyama, J., W. T. Platt, and V. B. Holland (1971). Deep-body temperature changes in rats exposed to chronic centrifugation. *Am. J. Physiol.*, 221:1271–7.

Pace, N., and D. F. Rahlman (1983). Thermoneutral zone and scaling of metabolic rate on body mass in small mammals. *Physiologist* [Suppl.], 26:S51–2.

Pace, N., and A. H. Smith (1983). Scaling of metabolic rate on body mass in small mammals at 2.0 *g*. *Physiologist* [Suppl.], 26:S125–6.

Panuska, J. A., J. M. Kilcoyne, and M. T. Fleming (1969). Righting reflexes and spontaneous locomotor activity during hypothermia. *Cryobiology*, 6:37–41.

Papanek, P. E., C. E. Wood, and M. J. Fregly (1991). Role of the sympathetic nervous system in cold-induced hypertension in rats. *J. Appl. Physiol.*, 71:300–6.

Paré, W. P. (1989). Age differences in the body temperature response to restraint–cold stress. *Physiol. Behav.*, 45:151–4.

Park, I. R., and J. Himms-Hagen (1988). Neural influences on trophic changes in brown adipose tissue during cold acclimation. *Am. J. Physiol.*, 255:R874–81.

Park, I. R., J. Himms-Hagen, and D. V. Coscina (1986). Long-term effects of lateral hypothalamic lesions on brown adipose tissue in rats. *Brain Res. Bull.*, 17:643–51.

Pasquis, P., A. Lacaisse, and P. Dejours (1970). Maximal oxygen uptake in four species of small mammals. *Respir. Physiol.*, 9:298–309.

Pendergrass, M., and D. D. Thiessen (1983). Sandbathing is thermoregulatory in the Mongolian gerbil, *Meriones unguiculatus*. *Behav. Neur. Biol.*, 37:125–33.

Pennycuik, P. R. (1969). Physiological and morphological characteristics of mice moved to 21°C after 10 generations at 34°C. *Aust. J. Biol. Sci.*, 22:677–87.

Pickard, G. E., R. Kahn, and R. Silver (1984). Splitting of the circadian rhythm of body temperature in the golden hamster. *Physiol. Behav.*, 32:763–6.

Pierau, F.-K., T. Tamasato, A. Cost, and S. Berkes (1984). Processing of afferent temperature signals in sensory ganglia and the spinal cord. *J. Therm. Biol.*, 9:51–5.

Pierau, F.-K., P. Torrey, and D. O. Carpenter (1975). Afferent nerve fiber activity responding to temperature changes of scrotal skin of the rat. *J. Neurophysiol.*, 38:601–12.

Pittman, Q. J., T. J. Malkinson, N. W. Kasting, and W. L. Veale (1988). Enhanced fever following castration: possible involvement of brain arginine vasopressin. *Am. J. Physiol.*, 254:R513–17.

Pittman, Q. J., and J. A. Thornhill (1990). Neuropeptide mechanisms affecting temperature control. *Curr. Top. Neuroendocrinol.*, 10:223–41.

Pohl, H. (1965). Temperature regulation and cold acclimation in the golden hamster. *J. Appl. Physiol.*, 20:405–10.

Poole, S., and J. D. Stephenson (1977a). Body temperature regulation and thermoneutrality in rats. *Q. J. Exp. Physiol.*, 143–9.

   (1977b). Core temperature: some shortcomings of rectal temperature measurements. *Physiol. Behav.*, 18:203–5.

Popovic, V. (1960). Survival time of hypothermic white rats (15°C) and ground squirrels (10°C). *Am. J. Physiol.*, 199:463–6.

Pospisilova, D., and L. Janský (1976). Effects of various adaptational temperatures on oxidative capacity of the brown adipose tissue. *Physiologica Bohem.*, 25:519–27.

Power, G. G. (1989). Biology of temperature: the mammalian fetus. *J. Dev. Physiol.*, 12:295–304.

Prange, A. J., Jr., C. B. Nemeroff, G. Bissette, P. J. Manberg, A. J. Osbahr, G. B. Burnett, P. T. Loosen, and G. W. Kraemer (1979). Neurotensin: distribution of hypothermic response in mammalian and submammalian vertebrates. *Pharmacol. Biochem. Behav.*, 11:473–7.

Prosser, C. L. (1973). Temperature. In C. L. Prosser (ed.), *Comparative Animal Physiology*, 3rd ed. (pp. 362–428). Philadelphia: Saunders.

(1986). *Adaptational Biology: Molecules to Organisms.* New York: Wiley.

Prosser, C. L., and J. E. Heath (1991). Temperature. In C. L. Prosser (ed.), *Comparative Animal Physiology, Environmental and Metabolic Animal Physiology,* 4th ed. (pp. 109–65). New York: Wiley-Liss.

Prychodko, W. (1958). Effect of aggregation of laboratory mice (*Mus musculus*) on food intake at different temperatures. *Ecology,* 39:500–3.

Puerta, M. L., M. P. Nava, M. Abelenda, and A. Fernandez (1990). Inactivation of brown adipose tissue thermogenesis by oestradiol treatment in cold-acclimated rats. *Pflügers Arch.,* 416:659–62.

Quan, N., and C. M. Blatteis (1989). Intrapreoptically microdialyzed and microinjected norepinephrine evokes different thermal responses. *Am. J. Physiol.,* 257:R816–21.

Quek, V., and P. Trayhurn (1990). Calorimetric study of the energetics of pregnancy in golden hamsters. *Am. J. Physiol.,* 259:R807–12.

Quintanar-Stephano, J. L., A. Quintanar-Stephano, and L. Castillo-Hernandez (1991). Effects of exposure to chronic-intermittent cold on thyrotropin and thyroid hormones in the rat. *Crybiology,* 28:400–3.

Raman, E. R., M. F. Roberts, and V. J. Vanhuyse (1983). Body temperature control of rat tail blood flow. *Am. J. Physiol.,* 245:R426–32.

Rampone, A. J., and M. E. Shirasu (1964). Temperature changes in the rat in response to feeding. *Science,* 144:317–19.

Rand, R. P., A. C. Burton, and T. Ing (1965). The tail of the rat, in temperature regulation and acclimatization. *Can. J. Physiol. Pharmacol.,* 43:257–67.

Randall, J. A., and D. D. Thiessen (1980). Seasonal activity and thermoregulation in *Meriones unguiculatus:* a gerbil's choice. *Behav. Ecol. Sociobiol.,* 7:267–72.

Ray, D. E., C. B. Roubicek, and M. Hamidi (1968). Organ and gland weights of rats chronically exposed to 22° and 35°C. *Growth,* 32:1–12.

Refinetti, R. (1989). Body size and metabolic rate in the laboratory rat. *Exp. Biol.,* 48:291–4.

(1990a). International directory of thermal biologists. *J. Therm. Biol.,* 15:183–92.

(1990b). Peripheral nervous control of cold-induced reduction in the respiratory quotient of the rat. *Int. J. Biometeor.,* 34:24–7.

Refinetti, R., and H. J. Carlisle (1986a). Complementary nature of heat production and heat intake during behavioral thermoregulation in the rat. *Behav. Neur. Biol.,* 46:64–70.

(1986b). Effects of lateral hypothalamic lesions on thermoregulation in the rat. *Physiol. Behav.,* 38:219–28.

Refinetti, R., and S. M. Horvath (1989). Thermopreferendum of the rat: inter- and intra-subject variability. *Behav. Neur. Biol.,* 52:87–94.

Refinetti, R., H. Ma, and E. Satinoff (1990). Body temperature rhythms, cold tolerance, and fever in young and old rats of both genders. *Exp. Gerontol.,* 25:533–43.

Refinetti, R., and M. Menaker (1992). The circadian rhythm of body temperature. *Physiol. Behav.,* 51:613–37.

Reigle, T. G., and H. H. Wolfe (1974). Potential neurotransmitters and receptor mechanisms involved in the central control of body temperature in golden hamsters. *J. Pharmacol. Exp. Ther.,* 189:97–109.

Reiter, R. J., K. Li, A. Gonzalez-Brito, M. G. Tannenbaum, M. K. Vaughan, G. M. Vaughan, and M. A. Villanua (1988). Elevated environmental temperature alters the responses of the reproductive and thyroid axes of female Syrian hamsters to afternoon melatonin injections. *J. Pineal Res.*, 5:301–15.

Reynolds, M. A., D. K. Ingram, and M. Talan (1985). Relationship of body temperature stability to mortality in aging mice. *Mech. Ageing Dev.*, 30:143–52.

Richard, D., J. Arnold, and J. LeBlanc (1986). Energy balance in exercise-trained rats acclimated at two environmental temperatures. *J. Appl. Physiol.*, 60:1054–9.

Richards, S. A. (1968). Vagal control of thermal panting in mammals and birds. *J. Physiol.*, 199:89–101.

Richardson, D. R., H. Qing-Fu, and S. Shepherd (1991). Effects of invariant sympathetic activity on cutaneous circulatory responses to heat stress. *J. Appl. Physiol.*, 71:521–9.

Richardson, D. R., S. Shepherd, and T. McSorley (1988). Evaluation of the role of skin temperature in the response of cutaneous capillary blood flow to indirect heat. *Microcirc. Endoth. Lymphatics*, 4:447–67.

Ricquier, D. (1989). Molecular biology of brown adipose tissue. *Proc. Nutr. Soc.*, 48:183–7.

Ricquier, D., L. Casteilla, and F. Bouillaud (1991). Molecular studies of the uncoupling protein. *FASEB J.*, 5:2237–42.

Ricquier, D., G. Mory, and P. Hemon (1979). Changes induced by cold adaptation in the brown adipose tissue from several species of rodents, with special reference to the mitochondrial components. *Can. J. Biochem.*, 57:1262–6.

Roberts, S. B., and W. A. Coward (1985). The effects of lactation on the relationship between metabolic rate and ambient temperature in the rat. *Ann. Nutr. Metab.*, 29:19–22.

Roberts, W. R. (1988). Differential thermosensor control of thermoregulatory grooming, locomotion, and relaxed postural extension. *Ann. N.Y. Acad. Sci.*, 525:363–74.

Roberts, W. R., and J. R. Martin (1977). Effects of lesions in central thermosensitive areas on thermoregulatory responses in rat. *Physiol. Behav.*, 19:503–11.

Roberts, W. R., and R. D. Mooney (1974). Brain areas controlling thermoregulatory grooming, prone extension, locomotion, and tail vasodilation in rats. *J. Comp. Physiol. Psychol.*, 86:470–80.

Roberts, W. R., R. D. Mooney, and J. R. Martin (1974). Thermoregulatory behaviors of laboratory rodents. *J. Comp. Physiol. Psychol.*, 86:693–9.

Robinson, P. F. (1959). Metabolism of the gerbil, *Meriones unguiculatus. Science*, 130:502–3.

(1968). General aspects of physiology. In R. A. Hoffman, P. F. Robinson, and H. M. Magalhaes (eds.), *The Golden Hamster, Its Use in Medical Research* (pp. 111–18). Ames: Iowa State University Press.

Rodland, K. D., and F. R. Hainsworth (1974). Evaporative water loss and tissue dehydration of hamsters in the heat. *Comp. Biochem. Physiol.*, 49A:331–45.

Romm, E., and A. C. Collins (1987). Body temperature influences on ethanol elimination rate. *Alcohol*, 4:189–98.

Roper, T. J. (1976). Sex differences in circadian wheel running rhythms in the Mongolian gerbil. *Physiol. Behav.*, 17:549–51.

Rosow, C. E., J. M. Miller, E. W. Pelikan, and J. Conchin (1980). Opiates and thermoregulation in mice. I: Agonists. *J. Pharmacol. Exp. Ther.*, 213:273–83.

Roth, J., G. Merker, F. Nurnberger, B. Pauly, and E. Zeisberger (1990). Changes in physiological and neuroendocrine properties during thermal adaptation of golden hamsters (*Mesocricetus auratus*). *J. Comp. Physiol.*, 160B:153–9.

Roth, J., E. Zeisberger, and H. Schwandt (1988). Influence of increased catecholamine levels in blood plasma during cold-adaptation and intramuscular infusion on thresholds of thermoregulatory reactions in guinea-pigs. *J. Comp. Physiol.*, 157B:855–63.

Rothwell, N. J., M. E. Saville, and M. J. Stock (1982). Effects of feeding a "cafeteria" diet on energy balance and diet-induced thermogenesis in four strains of rat. *J. Nutr.*, 112:1515–24.

Rothwell, N. J., and M. J. Stock (1980). Similarities between cold- and diet-induced thermogenesis in the rat. *Can. J. Physiol. Pharmacol.*, 58:842–8.

(1989). Surgical removal of brown fat results in rapid and complete compensation by other depots. *Am. J. Physiol.*, 257:R253–8.

Rothwell, N. J., M. J. Stock, and D. Stribling (1990). Diet-induced thermogenesis. In E. Schönbaum and P. Lomax (eds.), *Thermoregulation: Physiology and Biochemistry* (pp. 309–26). New York: Pergamon Press.

Roussel, B., and J. Bittel (1979). Thermogenesis and thermolysis during sleeping and waking in the rat. *Pflügers Arch.*, 382:225–31.

Roussel, B., A. Dittmar, and G. Chouvet (1980). Internal temperature variations during the sleep–wake cycle in the rat. *Waking Sleep*, 4:63–75.

Rousset, B., M. Cure, D. Jordan, A. Kervran, H. Bornet, and R. Mornex (1984). Metabolic alterations induced by chronic heat exposure in the rat: the involvement of thyroid function. *Pflügers Arch.*, 401:64–70.

Rozman, K., and H. Greim (1986). Toxicity of 2,3,7,8-tetrachlorodibenzo-*p*-dioxin in cold-adapted rats. *Arch. Toxicol.*, 59:211–15.

Rudy, T. A., J. W. Williams, and T. L. Yaksh (1977). Antagonism by indomethacin of neurogenic hyperthermia produced by unilateral puncture of the anterior hypothalamic/preoptic region. *J. Physiol.*, 272:721–36.

Rusak, B., and I. Zucker (1979). Neural regulation of circadian rhythms. *Physiol. Rev.*, 59:449–526.

Saetta, M., A. Noworaj, and J. P. Mortola (1988). Cold exposure of the pregnant rat and neonatal respiration. *Exp. Biol.*, 47:177–81.

Sakurada, S., O. Shido, and T. Nagasaka (1991). Mechanism of vasoconstriction in the rat's tail when warmed locally. *J. Appl. Physiol.*, 71:1758–63.

Satinoff, E. (1964). Behavioral thermoregulation in response to local cooling of the rat brain. *Am. J. Physiol.*, 206:1389–94.

(1978). Neural organization and evolution of thermal regulation in mammals. *Science*, 201:16–22.

(1983). A reevaluation of the concept of the homeostatic organization of temperature regulation. In E. Satinoff and P. Teitelbaum (eds.), *Handbook of Behavioral Neurobiology. Vol. 6: Motivation* (pp. 443–72). New York: Plenum Press.

(1991). Developmental aspects of behavioral and reflexive thermoregulation. In H. N. Shair et al. (eds.), *Developmental Psychobiology: New Methods and Changing Concepts* (pp. 169–88). New York: Oxford University Press.

Satinoff, E., and R. Henderson (1977). Thermoregulatory behavior. In W. K. Konig and J. E. R. Staddon (eds.), *Handbook of Operant Behavior* (pp. 153–73). Englewood Cliffs: Prentice-Hall.

Satinoff, E., S. Kent, and M. Hurd (1991). Elevated body temperature in female rats after exercise. *Med. Sci. Sports. Exer.*, 23:1250–3.

Satinoff, E., J. Liran, and R. Clapman (1982). Aberrations of circadian body temperature rhythms in rats with medial preoptic lesions. *Am. J. Physiol.*, 242:R352–7.

Satinoff, E., and R. A. Prosser (1988). Suprachiasmatic nuclear lesions eliminate circadian rhythms of drinking and activity, but not of body temperature, in male rats. *J. Biol. Rhyth.*, 3:1–22.

Satinfoff, E., and J. Rutstein (1970). Behavioral thermoregulation in rats with anterior hypothalamic lesions. *J. Comp. Physiol. Psychol.*, 71:77–82.

Satinoff, E., D. Valentino, and P. Teitelbaum (1976). Thermoregulatory cold-defense deficits in rats with preoptic/anterior hypothalamic lesions. *Brain Res. Bull.*, 1:553–65.

Scales, W. E., and M. J. Kluger (1987). Effect of antipyretic drugs on circadian rhythm in body temperature of rats. *Am. J. Physiol.*, 253:R306–13.

Schechtman, O., M. J. Fregly, P. van Bergen, and P. E. Papanek (1991). Prevention of cold-induced increase in blood pressure of rats by captopril. *Hypertension*, 17:763–70.

Schechtman, O., P. E. Papanek, and M. J. Fregly (1990). Reversibility of cold-induced hypertension after removal of rats from cold. *Can. J. Physiol. Pharmacol.*, 68:830–5.

Schingnitz, G., and J. Werner (1979). Responses of thalamic neurons to thermal stimulation of the limbs, scrotum and tongue in the rat. *J. Therm. Biol.*, 5:53–61.

(1986). Significance of scrotal afferents within the general thermoafferent system. *J. Therm. Biol.*, 11:181–9.

Schmidek, W. R., K. Hoshino, M. Schmidek, and C. Timo-Iaria (1972). Influence of environmental temperature on the sleep–wakefulness cycle of the rat. *Physiol. Behav.*, 8:363–71.

Schmidek, W. R., K. E. Zachariassen, and H. T. Hammel (1983). Total calorimetric measurements in the rat: influences of the sleep–wakefulness cycle and of the environmental temperature. *Brain Res.*, 288:261–71.

Schmidt, I. (1984). Interaction of behavioural and autonomic thermoregulation. In J. R. S. Hales (ed.), *Thermal Physiology* (pp. 309–18). New York: Raven Press.

Schmidt, I., R. Kaul, and H. J. Carlisle (1984). Body temperature of huddling newborn Zucker rats. *Pflügers Arch.*, 401:418–20.

Schmidt, I., R. Kaul, and G. Heldmaier (1987). Thermoregulation and diurnal rhythms in 1-week-old rat pups. *Can. J. Physiol. Pharmacol.*, 65:1355–64.

Schmidt-Nielsen, K. (1964). *Desert Animals. Physiological Problems of Heat and Water.* New York: Oxford University Press.

(1975a). Scaling in biology: the consequences of size. *J. Exp. Biol.*, 194:287–308.

(1975b). *Animal Physiology. Adaptation and Environment.* Cambridge University Press.

(1984). *Scaling: Why Is Animal Size so Important?* Cambridge University Press.

Schmidt-Nielsen, K., B. Schmidt-Nielsen, S. A. Jarnum, and T. R. Houpt (1957). Body temperature of the camel and its relation to water economy. *Am. J. Physiol.*, 188:103–12.

Schneider, J. E., L. A. Palmer, and G. N. Wade (1986). Effects of estrous cycles and ovarian steroids on body weight and energy expenditure in Syrian hamsters. *Physiol. Behav.*, 38:119–26.

Schneider, J. E., and G. N. Wade (1991). Effects of ambient temperature and body fat content on maternal litter reduction in Syrian hamsters. *Physiol. Behav.*, 49:135–9.

Schoenfeld, T. A., and C. M. Leonard (1985). Behavioral development in the Syrian golden hamster. In H. I. Siegel (ed.), *The Hamster* (pp. 289–321). New York: Plenum Press.

Scholander, P. F., R. Hock, V. Walters, F. Johnson, and L. Irving (1950). Heat regulation in some arctic and tropical mammals and birds. *Biol. Bull.*, 99:237–58.

Schönbaum, E., and P. Lomax (eds.) (1990). *Thermoregulation: Physiology and Biochemistry.* New York: Pergamon Press.

Selker, R. G., E. Bova, M. Kristofik, E. Jones, D. Iannuzzi, A. Landay, and F. Taylor (1979). Effect of total body temperature on toxicity of 1,3-bis(2-chloroethyl)-1-nitrosourea (BCNU). *Neurosurgery*, 4:157–61.

Severinsen, T., and N. A. Øritsland (1991). Endotoxin induced prolonged fever in rats. *J. Therm. Biol.*, 16:167–71.

Shellock, F. G., and S. A. Rubin (1984). Temperature regulation during treadmill exercise in the rat. *J. Appl. Physiol.*, 57:1872–7.

Shepherd, R. E., and P. D. Gollnick (1976). Oxygen uptake of rats at different work intensities. *Pflügers Arch.*, 362:219–22.

Shibata, H., and T. Nagasaka (1982). Contribution of nonshivering thermogenesis to stress-induced hyperthermia in rats. *Jpn. J. Physiol.*, 32:991–5.

(1987). The effect of forced running on heat production in brown adipose tissue in rats. *Physiol. Behav.*, 39:377–80.

Shido, O. (1987). Day–night variation of thermoregulatory responses to intraperitoneal electric heating in rats. *J. Therm. Biol.*, 12:273–9.

Shido, O., and T. Nagasaka (1990a). Thermoregulatory responses to acute body heating in rats acclimated to continuous heat exposure. *J. Appl. Physiol.*, 68:59–65.

(1990b). Heat loss responses in rats acclimated to heat loaded intermittently. *J. Appl. Physiol.*, 68:66–70.

Shido, O., S. Sakurada, M. Tanabe, and T. Nagasaka (1991). Temperature regulation during acute heat loads in rats after short-term heat exposure. *J. Appl. Physiol.*, 71:2107–13.

Shido, O., Y. Sugano, and T. Nagasaka (1986). Circadian change of heat loss in response to change in core temperature in rats. *J. Therm. Biol.*, 11:199–202.

Shido, O., Y. Yoneda, and T. Nagasaka (1989). Changes in body temperature of rats acclimated to heat with different acclimation schedules. *J. Appl. Physiol.*, 67:2154–7.

Shimada, S. G., and J. T. Stitt (1983). Inhibition of shivering during restraint hypothermia. *Can. J. Physiol. Pharmacol.*, 61:977–82.

Shimuzu, Y., and M. Saito (1991). Activation of brown adipose tissue thermogenesis in recovery from anesthetic hypothermia in rats. *Am. J. Physiol.*, 261:R301–4.

Shiota, M., and S. Masumi (1988). Effect of norepinephrine on consumption of oxygen in perfused skeletal muscle from cold-exposed rats. *Am. J. Physiol.*, 254:E482–9.

Shum, A., G. E. Johnson, and K. V. Flattery (1969). Influence of ambient temperature on excretion of catecholamines and metabolites. *Am. J. Physiol.*, 216:1164–9.

Sichieri, R., and W. R. Schmidek (1984). Influence of ambient temperature on the sleep–wakefulness cycle in the golden hamster. *Physiol. Behav.*, 33:871–7.

Simek, V. (1976). Influence of single administration of different diets on the energy metabolism at temperatures of 10, 20 and 30°C in the golden hamster. *Physiologica Bohem.*, 25:251–3.

Simon, E. (1981). Effects of CNS temperature on generation and transmission of temperature signals in homeotherma. A common concept for mammalian and avian thermoregulation. *Pflügers Arch.*, 392:79–88.

Simpkins, J. W. (1984). Spontaneous skin flusing episode in the aging female rat. *Maturitas*, 6:269–78.

Simpson, C. W., and G. E. Resch (1985). ACh and 5-HT stimulated thermogenesis at different core temperatures in the He-cold hypothermic hamster. *Brain Res. Bull.*, 15:123–7.

Singer, R., C. T. Harker, A. J. Vander, and M. J. Kluger (1986). Hyperthermia induced by open-field stress is blocked by salicylate. *Physiol. Behav.*, 36:1179–82.

Sivadjian, M. J. (1975). Etude comparative de la fonction sudorale chez le rat et la souris. *C. R. Acad. Sci.*, 280:77–80.

Skala, J. P. (1984). Mechanisms of hormonal regulation in brown adipose tissue of developing rats. *Can. J. Biochem. Cell Biol.*, 62:637–47.

Smith, R. E., and J. C. Roberts (1964). Thermogenesis of brown adipose tissue in cold-acclimated rats. *Am. J. Physiol.*, 206:143–8.

Snyder, G. K., and J. R. Nestler (1990). Relationships between body temperature, thermal conductance, Q_{10} and energy metabolism during daily torpor and hibernation in rodents. *J. Comp. Physiol.*, 159B:667–75.

Sod-Moriah, U. A. (1971). Reproduction in the heat-acclimatized female rat as affected by high ambient temperature. *J. Reprod. Fertil.*, 26:209–18.

(1974). Intrascrotal temperature, testicular histology and fertility of heat-acclimatized rats. *J. Reprod. Fertil.*, 37:263–8.

Spector, N. H., J. R. Brobeck, and C. L. Hamilton (1968). Feeding and core temperature in albino rats: changes induced by preoptic heating and cooling. *Science*, 161:286–8.

Spencer, F., H. W. Shirer, and J. M. Yochim (1976). Core temperature in the female rat: effect of pinealectomy. *Am. J. Physiol.*, 231:355–60.

Spencer, R. L., V. J. Hruby, and T. F. Burks (1990). Alteration of thermoregulatory set point with opioid agonists. *J. Pharmacol. Exp. Ther.*, 252:696–705.

Spiers, D. E., and E. R. Adair (1986). Ontogeny of homeothermy in the immature rat: metabolic and thermal responses. *J. Appl. Physiol.*, 60:1190–7.

Spiers, D. E., C. C. Barney, and M. J. Fregly (1981). Thermoregulatory responses of tailed and tailless rats to isoproterenol. *Can. J. Physiol. Pharmacol.*, 59:847–52.

Spray, D. C. (1986). Cutaneous temperature receptors. *Annu. Rev. Physiol.*, 48:625–38.

Stankiewicz, D. (1974). The course of pregnancy and the development of foetuses of mice under the conditions of hypothermy and hibernation with hypothermy. *Folia Biol.*, 22:91–102.

Steffen, J. M., and X. J. Musacchia (1985). Glucocorticoids and hypothermia induction and survival in the rat. *Cryobiology*, 22:385–91.

Steffen, J. M., and J. C. Roberts (1977). Temperature acclimation in the Mongolian gerbil (*Meriones unguiculatus*): biochemical and organ weight changes. *Comp. Biochem. Physiol.*, 58B:237–42.

Stitt, J. T. (1985). Evidence for the involvement of the organum vasculosum laminae terminalis in the febrile response of rabbits and rats. *J. Physiol.*, 368:501–11.

(1986). Prostaglandin E as the neural mediator of the febrile response. *Yale J. Biol. Med.*, 59:137–49.

Stitt, J. T., S. G. Shimada, and H. A. Bernheim (1985). Comparison of febrile responsiveness of rats and rabbits to endogenous pyrogen. *J. Appl. Physiol.*, 59:1721–5.

Stock, M. J. (1975). An automatic, closed-circuit oxygen consumption apparatus for small animals. *J. Appl. Physiol.*, 39:849–50.

(1989). The role of brown adipose tissue in diet-induced thermogenesis. *Proc. Nutr. Soc.*, 48:189–96.

Stolwijk, J. A. J., and J. D. Hardy (1974). Regulation and control in physiology. In V. B. Mountcastle (ed.), *Medical Physiology*, vol. 2 (pp. 1343–58). St. Louis: C. V. Mosby.

Stoner, H. B. (1968). Metabolism, heat loss and toxic factors. Mechanisms of body temperature changes after burns and other injuries. *Ann. N.Y. Acad. Sci.*, 150:722–37.

(1969). Studies on the mechanism of shock: the impairment of thermoregulation by trauma. *Br. J. Exp. Pathol.*, 50:125–38.

(1971). Effect of injury on shivering thermogenesis in the rat. *J. Physiol.*, 214:599–615.

(1972). Effect of injury on the responses to thermal stimulation of the hypothalamus. *J. Appl. Physiol.*, 33:665–71.

Stricker, E. M., J. C. Everett, and R. E. A. Porter (1968). The regulation of body temperature by rats and mice in the heat: effects of desalivation and the presence of a water bath. *Comm. Behavior. Biol.*, 2A:113–19.

Stricker, E. M., and F. R. Hainsworth (1970). Evaporative cooling in the rat: effects of hypothalamic lesions and chorda tympani damage. *Can. J. Physiol. Pharmacol.*, 48:11–17.

(1971). Evaporative cooling in the rat: interaction with heat loss from the tail. *Q. J. Physiol.*, 56:231–41.

Stupfel, M., V. Gourlet, A. Perramon, and C. Lemercerre (1989). Ultradian and circadian CO_2 emission variations in nocturnal and diurnal animals exposed to a light stimulus. *Comp. Biochem. Physiol.*, 94A:415–25.

Sugano, Y. (1983). Heat balance of rats acclimated to diurnal 2-hour feeding. *Physiol. Behav.*, 30:289–93.

Sundin, U., D. Herron, and B. Cannon (1981). Brown fat thermoregulation in developing hamsters (*Mesocricetus auratus*): a GDP-binding study. *Biol. Neonate*, 39:141–9.

Swift, R., and R. M. Forbes (1939). The heat production of the fasting rat in relation to the environmental temperature. *J. Nutr.*, 18:307–18.

Szelenyi, Z., and S. Donhoffer (1978). The effect of cold exposure on cerebral blood flow and cerebral available oxygen (aO_2) in the rat and rabbit: thermoregulatory heat production by the brain and the possible role of neuroglia. *Acta Physiol. Acad. Sci. Hung.*, 52:391–402.

Szelenyi, Z., and P. Hinckel (1987). Changes in cold- and heat-defence following electrolytic lesions of raphe nuclei in the guinea-pig. *Pflügers Arch.*, 409:175–81.

Szymusiak, R., A. DeMory, E. M. Kittrell, and E. Satinoff (1985). Diurnal changes in thermoregulatory behavior in rats with medial preoptic lesions. *Am. J. Physiol.*, 249:R219–27.

Szymusiak, R., and E. Satinoff (1981). Maximal REM sleep time defines a narrower thermoneutral zone than does minimal metabolic rate. *Physiol. Behav.*, 26:687–90.

(1982). Acute thermoregulatory effects of unilateral electrolytic lesions of the medial and preoptic area in rats. *Physiol. Behav.*, 28:161–70.

Szymusiak, R., E. Satinoff, T. Schallert, and I. Q. Wishaw (1980). Brief skin temperature changes towards thermoneutrality trigger REM sleep in rats. *Physiol. Behav.*, 25:305–11.

Takano, N., M. Mohri, and T. Nagasaka (1979). Body temperature and oxygen consumption of newborn rats at various ambient temperatures. *Jpn. J. Physiol.*, 29:173–80.

Takehiro, T. N., K. Shida, and Y. C. Lin (1979). Effects of 2,4-dinitrophenol on the body temperature and cardiopulmonary functions in unanesthetized rats *Rattus rattus*. *J. Therm. Biol.*, 4:297–301.

Tal, E., and F. G. Sulman (1975). Dehydroepiandrosterone-induced thyrotropin release during heat stress in rats. *J. Endocrinol.*, 67:99–103.

Talan, M. I. (1984). Body temperature of C57BL/6J mice with age. *Exp. Gerontol.*, 19:25–9.

Talan, M. I., and D. K. Ingram (1986a). Age comparisons of body temperature and cold tolerance among different strains of *Mus musculus. Mech. Ageing Dev.*, 33:247–56.

(1986b). Effects of voluntary and forced exercise on thermoregulation and survival in aged C57BL/6J mice. *Mech. Ageing Dev.*, 36:269–79.

Tanaka, H., K. Kanosue, T. Nakayama, and Z. Shen (1986). Grooming, body extension, and vasomotor responses induced by hypothalamic warming at different ambient temperatures in rats. *Physiol. Behav.*, 38:145–51.

Tanaka, J., T. Sato, I. K. Berezesky, R. T. Jones, B. F. Trump, and R. A. Cowley (1983). Effect of hypothermia on survival time and ECG in rats with acute blood loss. *Adv. Shock Res.*, 9:219–32.

Tarkkonen, H., and H. Julku (1968). Brown adipose tissue in young mice: activity and role in thermoregulation. *Experientia*, 24:798–9.

Taylor, D. C. M. (1982). The effects of nucleus raphe magnus lesions on an ascending thermal pathway in the rat. *J. Physiol.*, 326:309–18.

Taylor, P. M. (1960). Oxygen consumption in new-born rats. *J. Physiol.*, 154:153–68.

Tegowska, E. (1991). Stabilization of the brain temperature in mammals of different body size under various ambient temperatures. *Acta Theriologica*, 36:179–86.

Tempel, G. E., and L. H. Parks (1982). Brain norepinephrine and serotonin in the golden hamster during heat and cold acclimation and hypothermia. *Comp. Biochem. Physiol.*, 73C:377–81.

Tennent, D. M. (1946). A study of the water losses through the skin in the rat. *Am. J. Physiol.*, 145:436–40.

Thiessen, D. D. (1988). Body temperature and grooming in the Mongolian gerbil. *Ann. N.Y. Acad. Sci.*, 525:27–39.

Thiessen, D. D., C. Akins, and C. Zalaquett (1991). Exposure to odors from stressed conspecifics increases preference for higher ambient temperatures in C57BL/6J mice. *J. Chem. Ecol.*, 17:1611–19.

Thiessen, D. D., M. Graham, J. Perkins, and S. Marcks (1977). Temperature regulation and social grooming in the Mongolian gerbil (*Meriones unguiculatus*). *Behav. Biol.*, 19:279–88.

Thiessen, D. D., and E. M. Kittrell (1980). The harderian gland and thermoregulation in the gerbil (*Meriones unguiculatus*). *Physiol. Behav.*, 24:417–24.

Thomas, M. P., S. M. Martin, and J. M. Horowitz (1986). Temperature effects on evoked potential of hippocampal slices from noncold-acclimated, cold-acclimated and hibernating hamsters. *J. Therm. Biol.*, 11:213–18.

Thomasi, T. E., and B. A. Horwitz (1987). Thyroid function and cold acclimation in the hamster, *Mesocricetus auratus*. *Am. J. Physiol.*, 252:E260–7.

Thompson, G. E., and R. E. Moore (1968). A study of newborn rats exposed to cold. *Can. J. Physiol. Pharmacol.*, 46:865–71.

Thompson , G. E., and J. A. F. Stevenson (1965). A sex difference in the temperature response of rats to exercise. *Can. J. Physiol. Pharmacol.*, 43:437–43.

Thorington, R. W. (1966). The biology of rodent tails. A study of form and function. AAL-TR-65-8 (137 p.). Arctic Aeromedical Laboratory, Fort Wainwright, Alaska.

Thorne, P. S., C. P. Yeske, and M. H. Karol (1987). Monitoring guinea pig core temperature by telemetry during inhalation exposures. *Fund. Appl. Toxicol.*, 9:398–408.

Thornhill, J., and I. Halvorson (1990). Brown adipose tissue thermogenic responses of rats induced by central stimulation: effect of age and cold acclimation. *J. Physiol.*, 426:317–33.

(1992). Differences in brown adipose tissue thermogenic responses between Long-Evans and Sprague-Dawley rats. *Am. J. Physiol.*, 263:R59–69.

Thornhill, J., M. Hirst, and C. W. Gowdey (1978). Measurement of diurnal core temperatures of rats in operant cages by AM telemetry. *Can. J. Physiol. Pharmacol.*, 56:1047–50.

Tocco-Bradley, R., M. J. Kluger, and C. A. Kauffman (1985). Effect of age on fever and acute-phase response of rats to endotoxin and *Salmonella typhimurium*. *Infect. Immun.*, 47:106–11.

Tomasi, T. E., and B. A. Horwitz (1987). Thyroid function and cold acclimation in the hamster, *Mesocricetus auratus*. *Am. J. Physiol.*, 252:E260–7.

Toth, D. M. (1973). Temperature regulation and salivation following preoptic lesions in the rat. *J. Comp. Physiol. Psychol.*, 82:480–8.

Trayhurn, P. (1983). Decreased capacity for non-shivering thermogenesis during lactation in mice. *Pflügers Arch.*, 398:264–5.

(1985). Brown adipose tissue thermogenesis and the energetics of lactation in rodents. *Int. J. Obes.* [Suppl. 2], 9:81–8.

(1989). Brown adipose tissue and nutritional energetics – Where are we now? *Proc. Nutr. Soc.*, 48:165–75.

Trayhurn, P., and J. B. Douglas (1984). Fatty acid synthesis in brown adipose tissue of the Mongolian gerbil (*Meriones unguiculatus*): influence of acclimation temperature on synthesis in brown adipose tissue and the liver in relation to whole-body synthesis. *Comp. Biochem. Physiol.*, 78B:601–7.

Trayhurn, P., and W. P. James (1978). Thermoregulation and non-shivering thermogenesis in the genetically obese (ob/ob) mouse. *Pflügers Arch.*, 373:189–93.

Trayhurn, P., and D. G. Nicholls (1986). *Brown Adipose Tissue*. London: Edward Arnold.

Trayhurn, P., D. Richard, G. Jennings, and M. Ashwell (1983). Adaptive changes in the concentration of the mitochondrial "uncoupling" protein in brown adipose tissue of hamsters acclimated at different temperatures. *Biosci. Rep.*, 3:1077–84.

Trayhurn, P., and M. C. Wusteman (1987). Apparent dissociation between sympathetic activity and brown adipose tissue thermogenesis during pregnancy and lactation in golden hamsters. *Can. J. Physiol. Pharmacol.*, 65:2396–9.

Tulp, O. L., M. H. Gregory, and E. D. Danforth, Jr. (1982). Characteristics of diet-induced brown adipose tissue growth and thermogenesis in rats. *Life Sci.*, 30:1525–30.

Uehara, A., Y. Habara, and A. Kuroshima (1986). Effect of cold acclimation on glucagon receptors of rat white adipocytes. *Jpn. J. Physiol.*, 36:891–903.

Umpierre, C. C., and W. R. Dukelow (1978). Environmental heat stress effects in the hamster. *Teratology*, 16:155–8.

Unger, H., F. G. Hempel, and P. G. Kaufmann (1980). Correspondence of brain and rectal temperatures of guinea pigs in helium environments. *Undersea Biomed. Res.*, 7:27–34.

Vallerand, A. L., F. Perusse, and L. J. Bukowiecki (1990). Stimulatory effects of cold exposure and cold acclimation on glucose uptake in rat peripheral tissues. *Am. J. Physiol.*, 259:R1043–9.

Van Zoeren, J. G., and E. M. Stricker (1976). Thermal homeostasis in rats after intrahypothalamic injections of 6-hydroxydopamine. *Am. J. Physiol.*, 230:932–9.

Várnai, H., and S. Donhoffer (1970). Thermoregulatory heat production and regulation of body temperature in the new-born rat. *Acta Physiol. Acad. Sci. Hung.*, 37:35–49.

Veale, W. L., K. E. Cooper, and W. D. Ruwe (1984). Vasopressin: its role in antipyresis and febrile convulsion. *Brain Res. Bull.*, 12:161–5.

Villarroya, F., A. Felipe, and T. Mampel (1987). Reduced noradrenaline turnover in brown adipose tissue of lactating rats. *Comp. Biochem. Physiol.*, 86A:481–3.

Vinter, J., D. Hull, and M. C. Elphick (1982). Onset of thermogenesis in response to cold in newborn mice. *Biol. Neonate*, 42:145–51.

Vybíral, S., and J. F. Andrews (1979). The contribution of various organs to adrenaline stimulated thermogenesis. *J. Therm. Biol.*, 4:1–4.

Vybíral, S., and L. Janský (1974). Non-shivering thermogenesis in the golden hamster. *Physiologica Bohem.*, 23:235–43.

Wade, G. N., G. Jennings, and P. Trayhurn (1986). Energy balance and brown adipose tissue thermogenesis during pregnancy in Syrian hamsters. *Am. J. Physiol.*, 250:R845–50.

Wang, L. C. (1981). Effects of fasting on maximum thermogenesis in temperature-acclimated rats. *Int. J. Biometeor.*, 25:235–41.

Wang, L. C., and T. F. Lee (1990). Enhancement of maximal thermogenesis by reducing endogenous adenosine activity in the rat. *J. Appl. Physiol.*, 68: 580–5.

Wang, P. Y., D. W. Evans, N. Samji, and E. Llewellyn-Thomas (1980). A simple hygrometer for the measurement of evaporative loss from rodent skin. *J. Surg. Res.*, 28:182–7.

Watanabe, C., and T. Suzuki (1986). Sodium selenite-induced hypothermia in mice: indirect evidence for a neural effect. *Toxicol. Appl. Pharmacol.*, 86:373–9.

Watanabe, C., T. Suzuki, and N. Matsuo (1990a). Toxicity modification of sodium selenite by a brief exposure to heat or cold in mice. *Toxicology*, 64:245–53.

Watanabe, C., B. Weiss, C. Cox, and J. Ziriax (1990b). Modification by nickel of instrumental thermoregulatory behavior in rats. *Fund. Appl. Toxicol.*, 14:578–88.

Watkinson, W. P., J. W. Highfill, and C. J. Gordon (1989). Modulating effect of body temperature on the toxic response produced by the pesticide chlordimeform in rats. *J. Toxicol. Environ. Health*, 28:483–500.

Watkinson, W. P., M. J. Wiester, M. J. Campen, and V. M. Richardson (1992). Ozone toxicity in the unanesthetized, unrestrained rat: effect of changes in ambient temperature on physiological parameters. *The Toxicologist* (abstract), 12:230.

Weihe, W. H. (1965). Temperature and humidity climatograms for rats and mice. *Lab. Anim. Care*, 15:18–28.

(1973). The effect of temperature on the action of drugs. *Annu. Rev. Pharmacol.*, 20:409–25.

Weiss, B., and V. G. Laties (1961). Behavioral thermoregulation. *Science*, 133:1338–44.

Welch, W. R. (1980). Evaporative water loss from endotherms in thermally and hygrically complex environments: an empirical approach for interspecific comparisons. *J. Comp. Physiol.*, 139:135–43.

(1984). Temperature and humidity of expired air: interspecific comparisons and significance for loss of respiratory heat and water from endotherms. *Physiol. Zool.*, 57:366–75.

Wells, L. A. (1972). The effects of low body temperatures on deposition of tracers in the mammalian brain. *Cryobiology*, 9:367–82.

Werner, J., and A. Bienek (1990). Loss and restoration of preoptic thermoreactiveness after lesions of the rostral raphe nuclei. *Exp. Brain Res.*, 80:429–35.

Werner, J., G. Schingnitz, and J. Mathei (1986). Analysis of switching neurons within the thermoafferent system. *Exp. Brain Res.*, 64:70–6.

Werner, R. (1988). Effect of metopirone-ditartrate on thermogenesis in the guinea-pig. *Comp. Biochem. Physiol.*, 90C:445–50.

Weshler, Z., D. S. Kapp, P. F. Lord, and T. Hayes (1984). Development and decay of systemic thermotolerance in rats. *Cancer Res.*, 44:1347–51.

White, R. F., G. Feuerstein, and F. C. Barone (1992). Reduced brain temperature during transient focal ischemia is neuroprotective. *FASEB J.*, 6:A954 (abstract).

Whittow, G. C. (ed.) (1970). *Comparative Physiology of Temperature Regulation. Vol. 1: Invertebrates and Nonmammalian Vertebrates.* New York: Academic Press.

Wickler, S. J., B. A. Horwitz, S. F. Flaim, and K. F. LaNoue (1984). Isoproterenol-induced blood flow in rats acclimated to room temperature and cold. *Am. J. Physiol.*, 246:R747–52.

Wilkinson, C. W., H. J. Carlisle, and R. W. Reynolds (1980). Estrogenic effects on behavioral thermoregulation and body temperature of rats. *Physiol. Behav.*, 24:337–40.

Wilson, J. R., and L. M. Wilson (1978). Spontaneously hypertensive rat (SHR): possible elevation in thermoregulatory set-point. *Psychophysiology*, 15:295–6.

Wilson, L. M., P. J. Currie, and T. L. Gilson (1991). Thermal preference behavior in preweaning genetically obese (ob/ob) and lean (+/?, +/+) mice. *Physiol. Behav.*, 50:155–60.

Wilson, L. M., and H. L. Sinha (1985). Thermal preference behavior of genetically obese (ob/ob) and genetically lean (+/?) mice. *Physiol. Behav.*, 35:545–8.

Wilson, N. E., and E. M. Stricker (1979). Thermal homeostasis in pregnant rats during heat stress. *J. Comp. Physiol. Psychol.*, 93:585–94.

Wood, S. C. (1991). Interactions between hypoxia and hypothermia. *Annu. Rev. Physiol.*, 53:71–85.

Wood, S. C., J. W. Hicks, and R. K. Dupré (1987). Hypoxic reptiles: blood gases, body temperature and control of breathing. *Am. Zool.*, 27:27–9.

Wood, S. C., and G. M. Malvin (1991). Behavioral hypothermia: an adaptive stress response. In S. C. Wood, R. E. Weber, A. R. Hargens, and R. W. Millard (eds.), *Physiological Adaptations in Vertebrates: Respiration, Circulation, and Metabolism* (pp. 295–312). New York: Marcel Dekker.

Wright, G. L. (1976). Critical thermal maximum in mice. *J. Appl. Physiol.*, 40:683–7.

Wright, G. L., E. Knecht, D. Badger, S. Samueloff, M. Toraason, and F. Dukes-Dobos (1978a). Oxygen consumption in the spontaneously hypertensive rat. *Proc. Soc. Exp. Biol. Med.*, 159:449–52.

Wright, G., S. Iams, and E. Knecht (1977a). Resistance to heat stress in the spontaneously hypertensive rat. *Can. J. Physiol. Pharmacol.*, 55:975–82.

Wright, G., E. Knecht, and D. Wasserman (1977b). Colonic heating patterns and the variation of thermal resistance among rats. *J. Appl. Physiol.*, 43:59–64.

Wright, G., E. Knecht, and M. Toraason (1978b). Cardiovascular effects of whole-body heating in spontaneously hypertensive rats. *J. Appl. Physiol.*, 45:521–7.

Wünnenberg, W. (1983). Thermosensitivity of the preoptic region and the spinal cord in the golden hamster. *J. Therm. Biol.*, 8:381–4.

Wünnenberg, W., G. Kuhnen, and R. Laschefski-Sievers (1986). CNS regulation of body temperature in hibernators and non-hibernators. In H. C. Heller et al. (eds.), *Living in the Cold: Physiological and Biochemical Adaptations* (pp. 185–92). Amsterdam: Elsevier.

Wünnenberg, W., G. Merker, and E. Speulda (1976). Thermosensitivity of preoptic neurones in a hibernator (golden hamster) and a non-hibernator (guinea pig). *Pflügers Arch.*, 363:119–23.

Yahata, T., and A. Kuroshima (1987). Cold-induced changes in glucagon of brown adipose tissue. *Jpn. J. Physiol.*, 37:773–82.

(1989). Metabolic cold acclimation after repetitive intermittent cold exposure in rat. *Jpn. J. Physiol.*, 39:215–28.

Yamauchi, C., S. Fujita, T. Obara, and T. Ueda (1981). Effects of room temperature on reproduction, body and organ weights, food and water intake, and hematology in rats. *Lab. Anim. Sci.*, 31:251–8.

(1983). Effects of room temperature on reproduction, body and organ weights, food and water intakes, and hematology in mice. *Exp. Anim.*, 32:1–11.

Yanase, M., K. Kanosue, H. Yasuda, and H. Tanaka (1991). Salivary secretion and grooming behaviour during heat exposure in freely moving rats. *J. Physiol.*, 432:585–92.

Yanase, M., H. Tanaka, and T. Nakayama (1989). Effects of estrus cycle on thermoregulatory responses during exercise in rats. *Eur. J. Appl. Physiol.*, 58:446–51.

Yochim, J. M., and F. Spencer (1976). Core temperature in the female rat: effect of ovariectomy and induction of pseudopregnancy. *Am. J. Physiol.*, 231:361–5.

Young, A. A., and N. J. Dawson (1982). Evidence for on–off control of heat dissipation from the tail of the rat. *Can. J. Physiol. Pharmacol.*, 60:392–8.

(1988). Effects of environmental temperature on the development of a noradrenergic thermoregulatory mechanism in the rat. *Pflügers Arch.*, 412:141–6.

Yousef, M. K., and H. Johnson (1967). Time course of oxygen consumption in rats during sudden exposure to high environmental temperature. *Life Sci.*, 6:1221–8.

Zarrow, M. X., and M. E. Denison (1956). Sexual difference in the survival time of rats exposed to a low ambient temperature. *Am. J. Physiol.*, 186:216–18.

Zeisberger, E., and K. Brück (1971). Central effects of noradrenaline on the control of body temperature in the guinea-pig. *Pflügers Arch.*, 322:152–66.

Zeisberger, E., and J. Roth (1988). Role of catecholamines in thermoregulation of cold-adapted and newborn guinea pigs. In W. Kunzel and A. Jensen (eds.), *The Endocrine Control of the Fetus* (pp. 288–99). Berlin: Springer-Verlag.

Zeisberger, E., J. Roth, and E. Simon (1988). Changes in water balance and in release of arginine vasopressin during thermal adaptation in guinea-pigs. *Pflügers Arch.*, 412:285–91.

Index